Antimicrobial Stewardship for Nursing Practice

This book is enhanced with supplementary resources.
To access the customizable lecture slides please visit:
www.cabi.org/openresources/

Antimicrobial Stewardship for Nursing Practice

Edited by

Molly Courtenay

School of Healthcare Sciences, Cardiff University, UK

and

Enrique Castro-Sánchez

Imperial College London, UK

CABI is a trading name of CAB International

CABI
Nosworthy Way
Wallingford
Oxfordshire OX10 8DE
UK

CABI
745 Atlantic Avenue
8th Floor
Boston, MA 02111
USA

Tel: +44 (0)1491 832111
Fax: +44 (0)1491 833508
E-mail: info@cabi.org
Website: www.cabi.org

Tel: +1 (617)682-9015
E-mail: cabi-nao@cabi.org

British Library Cataloguing-in-Publication Data

A catalogue record for this book is available from the British Library, London.

Library of Congress Cataloging-in-Publication Data

Names: Courtenay, Molly, editor. | Castro-Sánchez, Enrique, editor.
Title: Antimicrobial stewardship for nursing practice / [edited by] Prof Molly Courtenay, Dr Enrique Castro-Sánchez.
Description: Wallingford, Oxfordshire ; Boston, MA : CABI, [2020] | Includes bibliographical references and index. | Summary: "With antimicrobial resistance on the rise, appropriate stewardship is more important than ever. This textbook, the first written by nurses for nurses, provides a clear and concise approach to good practice"-- Provided by publisher.
Identifiers: LCCN 2019036981 (print) | LCCN 2019036982 (ebook) | ISBN 9781789242690 (paperback) | ISBN 9781789242706 (epdf) | ISBN 9781789242713 (epub)
Subjects: MESH: Antimicrobial Stewardship | Nursing Process
Classification: LCC RT41 (print) | LCC RT41 (ebook) | NLM WB 330 | DDC 610.73--dc23
LC record available at https://lccn.loc.gov/2019036981
LC ebook record available at https://lccn.loc.gov/2019036982

ISBN: 9781789242690 (paperback)
      9781789242706 (ePDF)
      9781789242713 (ePub)

Commissioning Editor: Alexandra Lainsbury
Editorial Assistant: Lauren Davies
Production Editor: Shankari Wilford

Typeset by SPI, Pondicherry, India
Printed and bound in the UK by Bell & Bain Ltd, Glasgow, G46 7UQ

Contents

Contributors

Ligia Maria Abraão, RN, MSc, PhD, is Coordinator, Infection Control Service (Corporate Team), Americas Medical Services (United Health Group), São Paulo, Brazil.

Joanne Bosanquet, MBE, RN; RHV; BSc (Hons); MSc (Public Health); PG Dip; PD Cert; HonDUniv (Greenwich) is Chief Executive at the Foundation of Nursing Studies, London.

Emma Burnett, PhD, MSc, PGCert, BN, SPQ in IPC, RGN, is Associate Dean International & Academic Regional Lead for Middle East and North Africa, University of Dundee, Scotland.

Enrique Castro-Sánchez, PhD, MPH, BSc, RGN, DipTropNurs, PgDip, DLSHTM, FEANS, is Lead Academic Research Nurse, NIHR, HPRU in Healthcare Associated Infection and Antimicrobial Resistance at Imperial College, London, and Honorary Consultant Nurse in Communication & Patient Engagement, Imperial College Healthcare NHS Trust, London.

Molly Courtenay, PhD, MSc, BSc, RGN, PGCertEd, is Professor of Health Sciences, School of Healthcare Sciences, Cardiff University, Cardiff, UK. Email: courtenaym@cardiff.ac.uk

Briëtte du Toit, MSc.IPC, PgDiP.IPC, DipNurseSci, BCurNurs, is Infection Prevention and Control Officer, Mediclinic Southern Africa, Stellenbosch, South Africa.

Rose Gallagher, MBE, is Professional Lead Infection Prevention and Control/ AMR & Sustainability Lead, Royal College of Nursing, London.

Fiona Gotterson, RN, MN, MACN, is Project Manager and PhD Fellow at NCAS, Australia.

Nykoma Hamilton, RN, BN, PgCert, is Infection Prevention and Control Nurse, NHS Fife, Kirkcaldy, UK.

Heather Kennedy, MPharm, PgDip (Clin Pharm), PgDip (Infection Management), is an Advanced Antimicrobial Pharmacist, NHS Tayside, Dundee, UK.

Elizabeth Manias, RN, CertCritCare, BPharm, MPharm, MNurs, PhD, FACN (DLF), MPS, MSHPA, Board Certified Geriatric Pharmacist (BCGP), Accredited Pharmacist for Medication Management Reviews, is a Research Professor at the

School of Nursing and Midwifery, Faculty of Health, Deakin University, Centre for Quality and Patient Safety, Institute for Health Transformation, Australia. Email: emanias@deakin.edu.au

Jo McEwen, RGN, BSc, SpQ, MSc, NMP, is Advanced Nurse Practitioner, Antimicrobial Stewardship, NHS Tayside, Dundee, UK.

Rosely Moralez de Figueiredo, RN, MSc, PhD, is Professor, Nursing Department, Federal University of São Carlos, São Carlos-SP, Brazil.

Valerie Ness is Lecturer, School of Health and Life Sciences, Glasgow Caledonian University, Glasgow, UK.

Rita Olans, DNP, CPNP, APRN-BC, is Assistant Professor, School of Nursing, MGH Institute of Health Professions, Boston, Massachusetts.

Maria Clara Padoveze, RN, MSc, PhD, is Associate Professor, Department of Collective Health Nursing, School of Nursing, University of São Paulo, São Paulo-SP, Brazil.

Susie Singleton, RGN, PGDiP, MSc, is Consultant Nurse, Health Protection & Infection Prevention and Control, Public Health England, London.

Yolanda van Zyl, PgDiP.IPC, BCurNurs, is Infection Prevention and Control Specialist, Paarl Provincial Hospital, Paarl, South Africa.

Foreword

As the largest healthcare profession in the world, nursing is poised to make a significant impact on antimicrobial stewardship policy efforts both nationally and internationally. Nurses worldwide perform numerous functions that are critical to its success, such as prescribing antimicrobials, administering antimicrobial therapies, monitoring for adverse effects and documenting medication allergies.

Antimicrobial resistance, as a perspective of antimicrobial stewardship, is increasing globally and poses a serious threat for the populaces worldwide. In this view, nurses can play a greater role as educators to the correct and sustainable use of medicines and antibiotics, promoting public awareness of this phenomenon.

With these reflections in mind, educating undergraduate nurses for the role they will play in antimicrobial stewardship has been identified as a key activity for the containment of antimicrobial resistance by the World Health Organization (2016). However, there is a need to narrow the educational gap on these topics by providing nursing education based on an evidence-based framework of antimicrobial stewardship competencies achieving international consensus. This background, designed specifically for undergraduate nursing education programmes, will prepare the next generation of healthcare professionals for their advocacy role in antimicrobial therapy to ensure quality, safe patient care.

The authors blaze new trails in providing an international perspective on standardizing quality antimicrobial stewardship education and provide the foundation for transforming nursing knowledge, skills, attitudes and values that will shape the judgements essential for nurses and their interprofessional colleagues globally. Underpinned by Molly Courtenay's and Enrique Castro-Sánchez's international research, the book is written by nurses from across the globe for nurses but also for all healthcare professionals who can greatly take advantage of the evidence-based knowledge embedded in the various chapters of this textbook. Each chapter, based on one of the six core competencies, provides a succinct overview of the topic with evidence-based practice standards, case studies and a valuable reading list for furthering the development of clinical judgement and enhancing collaborative practice. Experienced nursing faculty and healthcare educators, for whom this topic may be a first endeavour, will find the book essential for developing new or enhancing existing antimicrobial stewardship curricula.

This comprehensive guide provides the tools to empower nursing faculty and healthcare professional programmes to standardize the complexities of antimicrobial stewardship and antimicrobial resistance education, prepare the next generation of nurses and healthcare professionals, and ultimately improve the outcomes for the patients they serve.

Celeste M. Alfes, DNP, MSN, RN, CNE, CHSE-A, FAAN
Associate Professor & Director of Nursing Education,
Simulation and Innovation
Frances Payne Bolton School of Nursing
Case Western Reserve University
Cleveland, Ohio
USA

Alessandro Stievano, PhD, MSN, RN, Ed.M, FAAN, FFNMRCSI
Professor, Health Professions Bachelor and Master's Degree Courses
University of Rome Tor Vergata, Italy
Research Coordinator, Centre of Excellence for
Nursing Scholarship, OPI of Rome
Italy

Introduction

Molly Courtenay*

School of Healthcare Sciences, Cardiff University, Cardiff, UK

Antimicrobial resistant infections, i.e. infections in which micro-organisms continue to grow in the presence of antimicrobials, cause approximately 700,000 deaths, globally, each year. This figure is predicted to rise to 10 million, combined with a cumulative cost of US$100 trillion, by 2050 (HM Government, 2019). The development of resistance is influenced by the frequency of antibiotic use (Holmes *et al.*, 2015). Antimicrobial stewardship (AMS) is described as 'a collection of coordinated interprofessional focused strategies to optimize antibiotic use by ensuring that every patient receives an antibiotic only when it is clinically indicated and then receives the appropriate antibiotic, at the right dose, duration and route of administration' (Manning *et al.*, 2015). Good AMS programmes should result in the best clinical outcome for the treatment or prevention of infection, with minimal toxicity to the patient and minimal impact on subsequent resistance (Gerding, 2001).

Nursing is the largest of the healthcare professions. Nurses are identified as important to AMS efforts in both international (Centers for Disease Control (CDC) and Prevention, 2019; American Association of Nurses (AAN), 2017; European Commission (EU), 2017; European Federation of Nurses (EFN) Associations, 2017) and national policy (National Health Services Scotland, 2014; Royal College of Nursing (RCN), 2018). Nurses, worldwide, perform numerous functions that are critical to the success of AMS programmes; not only are they increasingly prescribers of antimicrobials (AAN, 2017; Courtenay *et al.*, 2017), but they are also involved directly in a range of AMS activities (such as documentation of a medication allergy history, timely antibiotic administration, monitoring treatment and adverse events) (World Health Organisation (WHO), 2019). Furthermore, nurses are essential knowledge brokers (Broom *et al.*, 2016) at the centre of, and facilitators of, interprofessional collaborative practice essential for optimal AMS practice (Courtenay *et al.*, 2018). Nurses are well placed to educate the public and increase awareness of antimicrobial resistance (AMR) (RCN, 2014). For example, nurses working in general practice can communicate AMR messages to the general public. Health visitors can communicate

*Email: courtenaym@cardiff.ac.uk

these messages to young families, and school nurses are in an ideal position to communicate AMR messages to young people (RCN, 2014; EU, 2017; Reilly *et al.*, 2017).

Although the education of undergraduate healthcare professional students in AMS has been identified as a key activity for the containment of AMR (WHO, 2016), evidence suggests that only two thirds of nursing programmes incorporate any AMS teaching and only 12% cover all of the recommended AMS principles (Castro-Sánchez *et al.*, 2016).

Core AMS competencies exist for healthcare professional undergraduate education (Courtenay *et al.*, 2018) and these competencies have been used to achieve international consensus on a framework of AMS competencies (see Appendix 1) appropriate for nurse undergraduate education programmes (Courtenay *et al.*, 2019). This supports recommendations made by the International Council of Nurses (ICN) to include education about AMS in undergraduate nurse education programmes (ICN, 2017). The framework comprises overarching competency statements (subdivided into six domains) representing the knowledge, skills, attitudes and values that shape the judgements essential for AMS, and 63 individual competency descriptors, designed to reflect the level of experience of the learner and type of practice setting. The domains include infection prevention and control (18 competency descriptors), antimicrobials and AMR (5 competency descriptors), diagnosis of infection and use of antimicrobials (17 competency descriptors), antimicrobial prescribing practice (10 competency descriptors), person-centred care (7 competency descriptors) and interprofessional collaborative practice (6 competency descriptors). The competency framework supports standards of proficiency (Nursing and Midwifery Council (NMC), 2018), equipping newly registered nurses with the underpinning knowledge and skills required for their role in health promotion and protection and prevention of ill health.

Chapters 2–7 address each of the six domains, providing the necessary knowledge to enable students and practitioners to apply the competency descriptors in practice, no matter what their level of skill or type of practice setting/context is. Chapter 8 provides information for those leading AMS strategy with an understanding of how to implement AMS in nursing practice.

Appendix 1 – AMS Framework

DOMAIN 1: INFECTION PREVENTION AND CONTROL

COMPETENCY STATEMENT: All qualified healthcare professionals must understand the core knowledge underpinning infection prevention and control and use this knowledge appropriately to prevent the spread of infection.

Descriptors

To support antimicrobial stewardship, learners must demonstrate infection prevention and control by:

1. Describing what a micro-organism is;
2. Describing the different types of organisms that may cause infections;
3. Explaining what an antimicrobial resistant organism is;
4. Explaining the 'Chain of Infection';
5. Defining the components required for infection transmission, i.e. presence of an organism, route of transmission of the organism from one person to another (a host who is susceptible to infection);
6. Describing the routes of transmission of infectious organisms, i.e. contact, droplet, airborne routes;
7. Present and recognize the characteristics of a susceptible host;
8. Demonstrate an understanding of the importance of surveillance;
9. Describe how vaccines can prevent infections in susceptible persons;
10. Demonstrate the application of standard precautions in healthcare environments;
11. Apply appropriate policies/procedures and guidelines when collecting and handling specimens;
12. Apply policies, procedures and guidelines relevant to infection control when presented with infection prevention and control cases and situations;
13. Implement work practices that reduce the risk of infection (such as taking appropriate immunization or not coming to work when sick to ensure patient and other healthcare worker protection);
14. Appreciate that healthcare workers have the accountability and obligation to follow infection prevention and control protocols as part of their contract of employment;
15. Act as a role model to healthcare workers and members of the public by adhering to infection prevention and control principles;
16. Demonstrating knowledge and awareness of international/national strategies on infection prevention and control and antimicrobial resistance such as Global Action Plan for AMR and national recommendations, guidelines and legal requirements, or equivalent;
17. Understanding the role of the environment in optimal infection prevention and control practices, including hand hygiene and environmental cleaning;
18. Enabling infection prevention and control self-care for patients and families.

DOMAIN 2: ANTIMICROBIALS AND ANTIMICROBIAL RESISTANCE

COMPETENCY STATEMENT: All qualified healthcare professionals need to understand the core knowledge underpinning the concept of antimicrobial resistance and use this knowledge to help prevent antimicrobial resistance.

Descriptors

To support antimicrobial stewardship, learners must be able to:

1. Recognize the signs and symptoms of infection;
2. Discuss how inappropriate antimicrobial use (including non-adherence to treatment regime) may lead to antimicrobial resistance;

3. Identify approaches to support optimal prescribing of antimicrobials;

4. Recognize the importance of adequate specimen collection during relevant stages of antimicrobial use (i.e. prior to/during antibiotic treatment);

5. Describe how to recognize the appropriate response to antimicrobial treatment and the main signs that demonstrate antimicrobial failures.

DOMAIN 3: THE DIAGNOSIS OF INFECTION AND THE USE OF ANTIBIOTICS

COMPETENCY STATEMENT: All qualified healthcare professionals need to demonstrate knowledge in how infections are diagnosed and the appropriate use of antimicrobials, and use this knowledge appropriately to support the accurate diagnosis of infection and the appropriate use of antimicrobials.

Descriptors

To support antimicrobial stewardship, learners must be able to:

1. Explain how microbiology samples may aid diagnosis of infection;

2. Describe how to take, and demonstrate (following local procedures) the appropriate taking of, samples;

3. Interpret microbiology results/reports from the laboratory at a basic level;

4. Explain why self-limiting bacterial or viral infections are unlikely to benefit from antimicrobials;

5. Describe and demonstrate the self-management strategies required to treat self-limiting infections (i.e. analgesia/rest/fluids);

6. Understand the importance of following local antimicrobial policies (i.e. their development is based on local resistance patterns) and follow these policies in practice;

7. Explain the importance of documenting the indications for an antimicrobial (i.e. the route by which it is administered, its duration, dose, dose interval and review date) in clinical notes and demonstrate this in practice;

8. Demonstrate an understanding of the factors that need to be considered when choosing an antimicrobial (including site of infection and type of bacteria likely to cause an infection at a particular site);

9. Describe broad-spectrum and narrow-spectrum antimicrobials and the contribution of broad-spectrum antimicrobials to AMR;

10. Present and be able to recognize the common side effects associated with commonly administered antimicrobials;

11. Demonstrate an understanding of why documenting a patient allergy to an antimicrobial is important;

12. Explain why it is important to consider certain physiological conditions (such as renal function) in patients who receive an antimicrobial;

13. Describe what is meant by delayed prescribing;

14. Explain why it is essential that an accurate diagnosis of an allergy to an antimicrobial is based on history and laboratory tests;

15. Demonstrate an understanding of the role of the nurse regarding quality and safety of antibiotic prescriptions;

16. Demonstrate an awareness of laboratory results (i.e. culture and sensitivity) that demand prompt intervention;

17. Recognize antimicrobials that should be preserved for treatment of specific infections, e.g. carbapenemase-producing enterobacteriaceae (CPE) or colistin resistance or colistin-resistant pathogens.

DOMAIN 4: ANTIMICROBIAL PRESCRIBING PRACTICE

COMPETENCY STATEMENT: All qualified healthcare professionals need to be aware of how antimicrobials are used in practice in terms of their dose, timing, duration and appropriate route of administration, and apply this knowledge as part of their routine practice as follows:

Descriptors

To support antimicrobial stewardship, learners must be able to:

1. Explain how you would recognize and manage sepsis;

2. Describe why it is important to use local guidelines to initiate prompt, effective antimicrobial treatment in patients with life-threatening infections;

3. Describe why it is important to switch from intravenous antimicrobials to oral therapy;

4. Describe how to switch from IV antimicrobials to oral therapy;

5. Understand the appropriateness of antimicrobial administration models such as outpatient parenteral antimicrobial therapy (OPAT);

6. Demonstrate an understanding of the rationale and use of perioperative prophylactic antimicrobials to prevent surgical site infection;

7. Discuss factors that can influence antimicrobial prescribing and the implications for antimicrobial stewardship programmes;

8. Describe the national guidance on completion of a course of antimicrobials;

9. Explain how to identify the medicines with which antimicrobials can interact and why this is important;

10. Describe the difference between empiric, targeted and prophylactic antimicrobial therapy.

DOMAIN 5: PERSON-CENTRED CARE

COMPETENCY STATEMENT: All qualified healthcare professionals must seek out, integrate and value the input and engagement of the patient/carer as a partner in designing and implementing care.

Descriptors

To support antimicrobial stewardship that is patient-centred, learners need to:

1. Support participation of patients/carers as integral partners when planning/ delivering their care;

2. Share information with patients/carers in a respectful manner and in such a way that is understandable, encourages discussion and enhances participation in decision-making;

3. Ensure that appropriate education and support is provided by learners to patients/ carers and others involved with their care or service;

4. Listen respectfully to the expressed needs of all parties in shaping and delivering care or services;

5. Discuss patient/carer expectations or demands of antimicrobials and the need to use antimicrobials appropriately;

6. Recognize patient socio-economic restrictions (or other conditions of vulnerability) that may limit the appropriate course of antimicrobials, and support patients and their families for social protection achievement;

7. Recognize patients and families who require support to complete a course of antimicrobial therapy.

DOMAIN 6: INTERPROFESSIONAL COLLABORATIVE PRACTICE

COMPETENCY STATEMENT: All qualified healthcare professionals need to understand how different professions collaborate in relation to how they contribute to AMS.

Descriptors

To support AMS, learners are able to:

1. Demonstrate an understanding of the roles, responsibilities and competencies of other health professionals involved in antimicrobial treatment policy decisions;

2. Explain why it is important that healthcare professionals involved in the delivery of antimicrobial therapy (including the prescription, delivery and supply) have a common understanding of antimicrobial treatment policy decisions, the quantity of antimicrobial use and effective patient/client outcomes;

3. Establish collaborative communication principles and actively listen to other professionals and patients/carers involved in the delivery of antimicrobial therapy;

4. Communicate effectively to ensure common understanding of care decisions;

5. Develop trusting relationships with patients/carers and other health-/social care professionals;

6. Use information and communication technology effectively to improve interprofessional patient-centred care.

References

AAN (2017) White Paper: Redefining the antibiotic stewardship team: recommendations from the American Nurses Association/Centers for Disease Control and Prevention workgroup on the role of registered nurses in hospital antibiotic stewardship practices. Available at: https://www.cdc.gov/antibiotic-use/healthcare/pdfs/ANA-CDC-whitepaper.pdf (accessed 10 September 2019).

Broom, A., Broom, J., Kirby, E. and Scrambler, G. (2016) Nurses as antibiotic brokers. *Qualitative Health Research* 27(13), 1924–1935. Available at: https://doi.org/10.1177/1049732316679953 (accessed 10 September 2019).

Castro-Sánchez, E., Drumright, L.N., Gharbi, M., Farrell, S. and Holmes, A.H. (2016) Mapping antimicrobial stewardship in undergraduate medical, dental, pharmacy, nursing and veterinary education in the United Kingdom. *PLOS ONE* 11(2): e0150056. Available at: doi:10.1371/journal.pone.0150056 (accessed 10 September 2019).

CDC (Centers for Disease Control and Prevention) (2019) Core elements of hospital antibiotic stewardship programs. US Department of Health and Human Services, CDC, Altlanta, Georgia. Available at: https://www.cdc.gov/antibiotic-use/healthcare/pdfs/hospital-core-elements-H.pdf (accessed 20 December 2019).

Courtenay, M., Gillespie, D. and Lim, R. (2017) Patterns of dispensed non-medical prescriber prescriptions for antibiotics in primary care across England: a retrospective analysis. *Journal of Antimicrobial Chemotherapy* 72(10), 2915–2920.

Courtenay, M., Lim, R., Castro-Sánchez, E., Deslandes, R., Hodson, K., *et al.* (2018) Development of consensus-based national antimicrobial stewardship competencies for UK undergraduate healthcare professional education. *Journal of Hospital Infection* 100(3), 245–256.

Courtenay, M., Castro-Sánchez, E., Gallagher, R., McEwen, J., Bulabula, A.N.H., *et al.* (in press) Development of consensus based international antimicrobial stewardship competencies for undergraduate nurse education. *Journal of Hospital Infection*.

EFN (2017) Nurses are frontline combating antimicrobial resistance. Available at: http://www.efnweb.be/wp-content/uploads/EFN-AMR-Report-Nurses-are-frontline-combating-AMR-07-11-2017.pdf (accessed 10 September 2019).

EU (2017) *EU Guidelines for the Prudent Use of Antimicrobials in Human Health* (2017/C 212/01). Available at: https://ec.europa.eu/health/amr/sites/amr/files/amr_guidelines_prudent_use_en.pdf (accessed 10 September 2019).

Gerding, D.N. (2001) The search for good antimicrobial stewardship. *Joint Commission Journal on Quality and Patient Safety* 27(8), 403–404.

HM Government (2019) *Tackling Antimicrobial Resistance 2019–2024: The UK's Five-year National Action Plan*. Available at: https://assets.publishing.service.gov.uk/government/uploads/system/uploads/attachment_data/file/784894/UK_AMR_5_year_national_action_plan.pdf (accessed 10 September 2019).

Holmes, A.H., Moore, L.S.P., Sundsfjord, A., Stenbakk, M., Regmi, S., *et al.* (2015) Understanding the mechanisms and drivers of antimicrobial resistance. Available at: doi.org/10.1016/S0140-6736(15)00473-0 (accessed 10 September 2019).

International Council of Nurses (ICN) (2017) Position statement on antimicrobial resistance. ICN: Switzerland.

Manning, M.L., Pfeiffer, J. and Larson, E.L. (2015) Combating antibiotic resistance: the role of nursing in antibiotic stewardship. *American Journal of Infection Control* 44(12), 1454–1457.

National Health Services Scotland (2014) *Scottish Management of Antimicrobial Resistance Action Plan 2014–2018 (ScotMARAP 2)*. The Scottish Government, Edinburgh, UK.

Nursing and Midwifery Council (NMC) (2018) Future Nurse: Standards of proficiency for registered nurses. NMC, London.

RCN (Royal College of Nursing) (2014) *Defining nursing*. Royal College of Nursing, London.

RCN (2018) Antimicrobial resistance. Available at: https://www.rcn.org.uk/clinical-topics/infection-prevention-and-control/antimicrobial-resistance (accessed 10 September 2019).

Reilly, J., Ness, V. and MacDonald, E. (2017) The health visitor role in containing antimicrobial resistance. *Journal of Health Visiting* 5(8), 386.

WHO (World Health Organization) (2016) *Nursing Midwifery Services: Strategic Directions 2011–2015*. Available at: http://apps.who.int/iris/bitstream/10665/70526/1/WHO_HRH_HPN_10.1_eng.pdf (accessed 21 March 2016).

WHO (2019) *Competency Framework for Health Workers' Education and Training on Antimicrobial Resistance*. World Health Organization, Geneva, Switzerland. Availablein:http://apps.who.int/medicinedocs/documents/s23443en/s23443en.pdf (accessed 10 September 2019).

Infection Prevention and Control   2

Briëtte du Toit[1],* and Yolanda van Zyl[2]

[1]Infection Prevention and Control Officer, Mediclinic Southern Africa, Stellenbosch, S.A.; [2]Infection Prevention and Control Specialist, Paarl Provincial Hospital, Paarl, S.A.

Objective: For the student to understand the core knowledge underpinning infection prevention and control, and how to apply this knowledge to nursing practice to prevent the spread of infection.

Micro-organisms

Micro-organisms (also called microbes), such as bacteria, fungi and viruses, are living organisms too small to be seen with the naked eye but visible under a microscope. Micro-organisms are the most frequent life form on earth, found in almost every conceivable environment and essential to sustaining life, e.g. breaking down dead animal and vegetal matter into simpler substances used at the beginning of the food chain. Micro-organisms are also used in the production of food and the breakdown of sewage and other toxic wastes into safe matter, e.g. fermentation (Sood, 2013).

There are five basic groups of micro-organisms:

- bacteria
- fungi
- viruses
- parasites
- algae.

(Sood, 2013)

Bacteria

Bacteria are, typically, single-celled (prokaryotic) organisms that reproduce by binary fission (by dividing into two halves), producing two new cells. These cells then divide again and again. They contain both deoxyribonucleic acid (DNA) and ribonucleic acid (RNA), but no defined nucleus. They usually have a cell wall and may possess other features such as pili, fimbriae and flagella (Fig. 2.1).

---

*Corresponding author: briette.dutoit@mediclinic.co.za

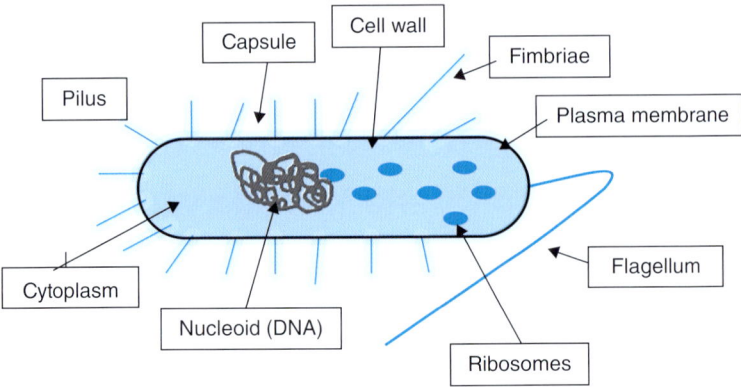

Fig. 2.1. Bacterial cell anatomy and internal structure.

Bacteria can be divided into groups based on the Gram stain characteristics and their shape. The Gram stain is a widely used technique to classify bacteria. Bacteria are classified as either Gram-positive (e.g. *Streptococci*) or Gram-negative (e.g. *Enterobacteriaceae*), based on the characteristics of their cell wall. Gram staining is used to colour the cells red or violet. Gram-positive bacteria stain violet during the decolouring process due to the presence of a thick layer of peptidoglycan in their cell walls, which retains the crystal violet when they are stained. Alternatively, Gram-negative bacteria stain red due to a thinner peptidoglycan wall, which does not retain the crystal violet during this process (Forder, 2005). Knowledge about the Gram stain is important as it guides the commencement of antibiotic treatment. Decisions to de-escalate treatment can then be made later based on the type of organism that is cultured and its sensitivity to antibiotics.

Based on their shape, bacteria can be further classified into the following categories (see Fig. 2.2):

- Cocci
    - Gram-positive, e.g. *Staphylococci*
    - Gram-negative, e.g. *Neisseria spp.*

- Bacilli
    - Gram-positive, e.g. *Clostridia, Bacillus spp.*
    - Gram-negative, e.g. *Escherichia coli, Pseudomonas spp.*

- Spiral or curved rods
    - Gram-negative, e.g. vibrios
    - Poorly Gram-negative, e.g. spirochaetes

Fungi

Fungi (yeasts and moulds) are, typically, single-celled, eukaryotic micro-organisms that reproduce asexually by budding. They possess RNA and DNA, a defined nucleus and a cell wall. Yeasts (e.g. *Candida* spp.) are small and round. Moulds (e.g. *Mucor* spp.) are, typically, filamentous and reproduce by asexual spores.

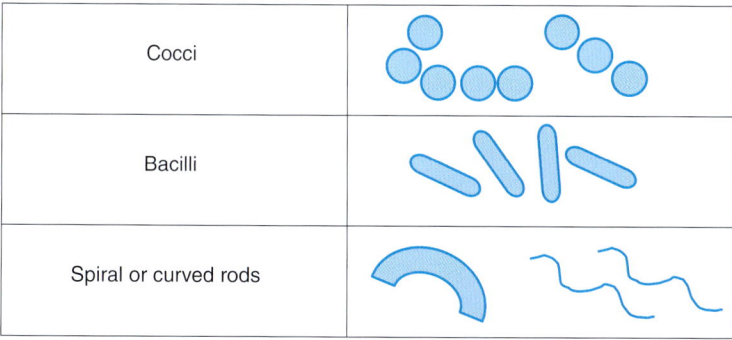

Cocci	
Bacilli	
Spiral or curved rods	

Fig. 2.2. Bacterial shapes and arrangements.

Viruses

Viruses are acellular, infectious particles that can only replicate inside a living host cell. Viruses are too small to be seen with an ordinary microscope. The vast majority of viruses possess either DNA or RNA (Sood, 2013). Several groups can be recognized:

- single-stranded RNA viruses, e.g. bunyaviruses, coronaviruses, retroviruses;
- double-stranded RNA viruses, e.g. reoviruses;
- segmented RNA viruses, e.g. arenaviruses.

(Sood, 2013)

Parasites

Parasites use a living host to survive and replicate. Parasites consist of protozoa and helminths. Protozoan parasites include:

- Sporozoa, e.g. *Plasmodium* sp., *Toxoplasma gondii*, *Cryptosporidium* sp.;
- rhizopoda, e.g. *Entamoeba* sp.;
- flagellates, intestinal, e.g. *Trichomonas* sp., blood, e.g. *Trypanosoma* sp., other, e.g. *Pneumocystis jirovecii*.

(Sood, 2013)

Requirements for the reproduction of microorganisms

When placed in a suitable nutritious environment, and maintained under appropriate conditions, a bacterial cell begins to grow. When it has made twice the amount of initial component materials, it will then divide. Requirements for microbial growth fall into two categories: physical and chemical (Sood, 2013). Physical requirements include temperature, pH and osmotic pressure. Chemical requirements include water, sources of carbon and nitrogen, minerals, oxygen and organic growth factors (Sood, 2013).

Indigenous human microbiota

The indigenous human body microbiota is the most important source of micro-organisms that can cause human disease. Examples of occupied habitats include our oral cavity, genital organs, respiratory tract, skin and gastrointestinal system (Lloyd-Price *et al.*, 2016). The human microbiota is estimated to be 10^{13}–10^{14} microbial cells, with around a ration of 1:1 microbial to human cells (Sender *et al.*, 2016). These numbers are derived from the total bacterial cells in the colon (3.8×10^{13} bacteria), the organ that harbours the highest number of microbes (Sender *et al.*, 2016). The gastrointestinal microbiota is predominantly composed of bacteria from three major phyla, namely *Firmicutes, Bacteroidetes* and *Actinobacteria* (Tap *et al.*, 2009).

Specific micro-organisms have specific body sites in which they live and multiply without causing disease. This is called colonization. Colonization is referred to as the presence of micro-organisms in or on a host, with growth and multiplication, but without tissue invasion or cellular injury (Public Health Agency of Canada (PHAC), 2013). A colonized person shows no obvious signs of disease yet can spread micro-organisms into the environment through normal day-to-day activities.

What is an antibiotic resistant organism?

Antibiotic resistance (see also Chapter 3) is a natural evolutive phenomenon which develops when micro-organisms adapt to, and grow in, the presence of antibiotics. The development of resistance is influenced by the frequency of antibiotic use (Holmes *et al.*, 2016). Bacteria can acquire various resistance mechanisms, resulting in resistance to several antimicrobial agents, and limit the future treatment options of bacterial infections (European Centre for Disease Control and Prevention (ECDC), 2018). Mechanisms of resistance against antimicrobial agents include decreased permeability of the bacterial cell wall, alteration of the drug target, inactivation of the drug, active efflux (mechanism) of the drug and development of alternative metabolic pathways (World Health Organization (WHO), 2015, p. 9). Drug-resistant bacteria can circulate and transfer from human beings and animals, through food, water and the environment (Singer *et al.*, 2016). Transmission of resistant bacteria is also influenced by trade, travel and both human and animal migration (Singer *et al.*, 2016). Resistant bacteria can be found in food animals and food products consumed by humans. In healthcare settings, resistant micro-organisms also spread through inadequate infection prevention and control programmes (ECDC, 2018).

Because many antibiotics belong to the same class, resistance to one specific antibiotic can lead to resistance to a whole related class. Resistance that develops in one organism can also spread rapidly to other organisms through, for instance, exchange of genetic material between different bacteria. The 'resistome', a term coined by Wright (2007), reflects the collection of all the antibiotic resistance genes and their precursors in both pathogenic and non-pathogenic bacteria. Causes of the global resistome are overpopulation, enhanced global migration, increased use of antibiotics in human healthcare, animal farming and food production, se-lection pressure, poor sanitation, wildlife spread (*E. coli* and enterococci in wild animals demonstrates that wildlife has the potential to serve as an environmental reservoir and melting-pot of bacterial resistance), and poor sewage and disposal systems (Singer *et al.*, 2016). Antimicrobial resistance (AMR) can further develop naturally through the mutations in bacterial genes and can be transmitted be-tween bacteria (ECDC, 2018).

Chain of Infection

The chain of infection includes six different links (see Fig. 2.3): pathogen (infec-tious agent), reservoir, portal of exit, mode of transmission, portal of entry and new susceptible host. Each link has a unique role in the chain, and each can be interrupted through various means.

Stages of infection

○   Acquisition of micro-organisms
    The initial point of contact with a given microbial species is of fundamental importance and can be acquired from microbiota (endogenous) or from ex-ternal sources (exogenous). Major routes of transmission of micro-organisms are:

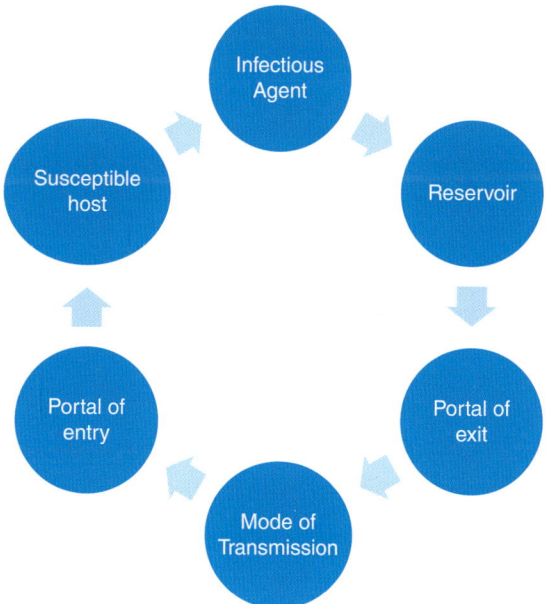

Fig. 2.3. Chain of infection.

– Respiratory
  • Airborne route (inhalation): particles (aerosols) <5 μm (measles, pulmonary tuberculosis)
  • Droplet route: particles >5 μm (influenza, pertussis)
– Contact
  • Direct contact, e.g. the hands of healthcare workers
  • Indirect contact, via the environment and contaminated equipment
– Vector/inoculation or trauma, e.g. malaria, tetanus
– Ingestion/faecal–oral route, e.g. gastroenteritis, hepatitis A infection
– Transplacentally (from a mother to her unborn baby), e.g. congenital toxoplasmosis

○ Colonization
Acquisition of a new microbial species establishes itself in its new habitat and multiplies under local conditions without invading the host and causing disease, e.g. *Staphylococcus aureus* in the anterior nares of the nose.

○ Penetration of anatomical barriers
In order to invade living human tissue, a micro-organism must breach the relevant surface barrier, e.g. *Acinetobacter baumanii* in an open wound.

○ Spread
An invading micro-organism may spread by one or more routes: directly through surrounding tissues, along tissue planes or via the veins and lymphatic vessels, e.g. bloodstream infection caused by *E. coli* from a bladder infection.

○ Mechanisms of damage by micro-organisms:
  ▪ Bulk effect – the sheer bulk of organisms may obstruct a hollow organ, e.g. some helminth infection of the intestine.
  ▪ Toxin mediated – the toxin released causes the main features of disease, e.g. *Clostridium tetani* releasing tetanospasmin (toxin).
  ▪ Altered function of host system, e.g. increased bowel motility leading to diarrhoea, or coughing and sneezing, e.g. *Haemophilus influenzae* causing upper respiratory disease.
  ▪ Host response to infection – inflammatory reaction that causes swelling, increased fragility of tissues, formation of pus, scaring or necrosis (Bauman, 2018).

Characteristics of the susceptible host

After an infectious agent penetrates the body, it has to multiply to cause a disease. In some hosts, infection leads to the development of symptoms and progression of the disease, but other hosts can behave differently. Individuals who are likely to develop a disease after exposure to an infectious agent are called susceptible hosts. Different individuals are not equally susceptible to infection, for a variety of reasons.

Factors that increase the susceptibility of a host to the development of a disease are called risk factors. Some risk factors arise from outside the individual – for example, poor personal hygiene (Bauman, 2018), or poor control of reservoirs of infection in the environment. Factors such as these increase the exposure of susceptible hosts to infectious agents, which makes the disease more likely to develop, e.g. a person contracting cholera by ingesting contaminated water (faecal–oral route) (WHO, 2015).

Additionally, some people in a community are more likely to develop a disease than others, even though they all may have been equally exposed to infectious agents. Such difference in response can be due to a low level of immunity within the more susceptible individuals. Immunity refers to the resistance of a person to communicable diseases, reflecting the effectiveness of white blood cells and antibodies (defensive proteins) to fight infectious agents. Low levels of immunity could be due to diseases like HIV/AIDS, which suppress immunity; poorly developed or immature immunity, as in very young children; not being vaccinated; poor nutritional status (e.g. malnourished children); and pregnancy (Kannan, 2016).

Surveillance

Describe the role of surveillance in antimicrobial resistance

Surveillance refers to the continuous and systematic collection and analysis of data to detect and monitor public health threats, including the epidemiology of these threats, the burden of disease and the identification of populations at high risk and age groups (CDC, 2018a; WHO, 2018). The collection of epidemiological data and their correlation with clinical and demographic data help to understand the burden of disease and the magnitude of a problem worldwide (Johnson, 2015). Surveillance data also provides information on the development and transmission of antimicrobial resistance (AMR) between humans and animals through food, water and the environment. This information enables clinicians, public health officers and researchers to identify and monitor emerging drug-resistant pathogens and their mechanisms of resistance. In turn, such knowledge would facilitate the development of new diagnostic tools and methods to detect AMR and measure the effectiveness of any interventions designed to reduce AMR (Johnson, 2015; WHO, 2015, 2017).

Surveillance plays a key role in understanding the epidemiology of resistance, and the factors influencing the development and transmission of resistance. Priority areas identified for AMR surveillance might differ, but the following are recommended (UKDOH, 2014; Johnson, 2015; WHO, 2018):

- Identify the key human and veterinary pathogens that require monitoring for resistance to critically important antimicrobials, as well as their prevalence and geographical patterns (UKDOH, 2014; Johnson, 2015; WHO, 2015, 2018).
- Monitor the usage of antimicrobials in humans and animals (UKDOH, 2014; Johnson, 2015; WHO, 2015, 2018).

- Document any unintended consequences of reduced antimicrobial prescribing (UKDOH, 2014; Johnson, 2015).
- Assess the knowledge and attitudes of the public towards appropriate antibiotic usage (CDC, 2013; UKDOH, 2014; Johnson, 2015; ECDC, 2018).
- Determine the level of professional engagement of healthcare providers by monitoring the uptake of critically important antibiotics (UKDOH, 2014; Johnson, 2015).
- Review the actions that were implemented as a result of international collaboration to reduce global transmission of AMR pathogens (CDC, 2013; UKDOH, 2014; Johnson, 2015; WHO, 2015, 2018; ECDC, 2018).

Surveillance programmes should meet the following requirements:

- Detect and monitor the prevalence, susceptibility and geographical patterns of pathogens and the emergence of AMR.
- Provide knowledge on how resistance develops and spreads, including the circulation of resistance between and within humans and animals through water, the environment and food.
- Collect, analyse and report data on the occurrence and trends of resistance in relevant pathogens and the identification of newly emerging resistance.
- Identify populations at risk.
- Monitor the use of antimicrobials in human and animal health, and agriculture.
- Use data to inform policy and guideline development, and formulation of interventions.
- Guide the formulation of interventions and assess the impact thereof.
- Monitor the prevalence of healthcare-associated infections and the susceptibility of pathogens causing these.
- Use surveillance data for the improvement of infection prevention and control (IPC) programmes (WHO, 2015; IACG, 2018, p. 1).

An important aspect of surveillance is the prompt feedback of findings to stakeholders, to ensure appropriate action is taken to reduce and prevent further transmission of diseases, and to assist with policy-making.

Vaccines

Describe how vaccines prevent infections in a susceptible host

Vaccination is the process of inducing immunity against a specific disease. Immunity can be induced either passively or actively. Passive immunity is generated through the administration of an antibody-containing preparation. Active immunity is achieved by administering a vaccine or toxoid to stimulate the immune system to produce a prolonged humoral and/or cellular immune response.

(Kliegman *et al.*, 2019; CDC [Vaccines: the basics])

Table 2.1. List of available vaccines. (From CDC, 2016a)

Diseases			
Adenovirus	*Haemophilus influenza type b* (HIB)	Mumps	Rubella
Anthrax	Human papillomavirus (HPV)	Pertussis	Shingles
Cholera	Seasonal influenza	Pneumococcal	Smallpox
Diphtheria	Japanese encephalitis	Polio	Tetanus
Hepatitis A	Measles	Rabies	Tuberculosis
Hepatitis B	Meningococcal	Rotavirus	Typhoid fever
		Varicella	Yellow fever

There is currently a wide range of vaccines available worldwide and countries have their own specific vaccines and regimes. Table 2.1 provides an overview of some of the diseases for which vaccines are available.

Vaccination is one of the most effective public health interventions to prevent infection in humans and animals. Vaccines are a cost-effective and significant intervention to prevent the onset and transmission of infectious diseases and can further reduce the development of resistance by minimizing the requirement for antimicrobials (Lipsitch and Siber, 2016; EC, 2017).

Vaccines reduce the burden of infectious diseases worldwide. They have the ability to prevent childhood infections and decrease morbidity and mortality. Vaccines protect through direct immunization, by activating the immune system to identify and react to an infectious agent, preventing an infection or decreasing the magnitude of the disease. Vaccines can further protect unvaccinated individuals who cannot be vaccinated (e.g. individuals with some chronic illness) or are yet to be vaccinated (e.g. newborn babies) by a process called herd immunity or community protection, extending protection to unvaccinated persons in the population. (Phizer, 2014; Lipsitch and Siber, 2016; Jansen and Anderson, 2018). Herd immunity refers to the effect of a successful vaccine programme in which enough people within a community are immune to an infectious agent due to vaccination and/or previous illness, which in turn reduces the transmission of such specific pathogen from one person to the other. The protective effect of herd immunity increases with the number of vaccinated individuals (CDC, 2016a). Vaccines help prevent the transmission of infectious diseases in asymptomatic carriers, averting disease and subsequent antimicrobial usage (WHO, 2016; Chisholm *et al.*, 2018).

Vaccines also contribute towards antimicrobial stewardship efforts by:

- preventing or limiting infectious diseases, thus reducing the use of antimicrobials (WHO, 2015; Klugman and Black, 2018);
- decreasing the prevalence of viral infections, which are often inappropriately treated with antibiotics (WHO, 2015);
- mitigating the risk of secondary infections that may require antibiotic treatment (Klugman and Black, 2018);

- preventing diseases that are difficult to treat or untreatable due to antimicrobial resistance (WHO, 2015); and
- reducing infections in food animals (WHO, 2015).

Collecting and Handling Specimens

Procedures

Unnecessary and clinically questionable laboratory tests may harm patients through exposure to avoidable medical treatment and inappropriate antibiotics treatment, also increasing the cost of healthcare delivery (MacVane *et al.*, 2016, pp. 3–4).

Specimens for microbiological culture (see also Chapter 4) should be collected as soon as possible after the admission of a patient to ensure accurate diagnosis and appropriate antimicrobial therapy. The correct management of specimens is also crucial to maximize the cost-effectiveness and clinical relevance of microbiological testing. Clinicians need assurance that laboratory results are accurate, significant and clinically relevant. If specimen collection and management are not a priority, the laboratory can contribute little or nothing to ensure safe and effective patient care. Reporting of accurate, but insignificant, information can be as harmful as reporting incorrect results. The correct interpretation of the laboratory report is important to ensure optimal treatment. Healthcare workers should receive training on their interpretation. The presence of a micro-organism does not necessarily indicate an infection. This highlights the importance of taking clinical signs and symptoms into consideration when a diagnosis is made and treatment prescribed (Mediclinic Southern Africa, 2015).

Appropriate management of patient care and the application of the correct procedures when collecting a specimen are key to accurate laboratory diagnosis. The laboratory diagnosis directly affects patient care and patient outcomes, influences therapeutic decisions and impacts on infection prevention and control, length of hospital stay, costs of treatment and laboratory cost and efficiency (Baron *et al.*, 2013).

Important considerations when collecting a specimen include:

- maintenance of standards throughout the process of collection, transport and storage (Baron *et al.*, 2013);
- ensuring that specimens are collected prior to the administration of antibiotics – the microbiota changes after commencement of antibiotics, potentially impacting on the reliability of results (Baron *et al.*, 2013); however, collection should not delay antimicrobial administration in critically ill patients;
- prevention of contamination of the specimen with commensal bacteria present in body sites, e.g. lower respiratory tract (sputum) (Baron *et al.*, 2013);
- avoidance of routine collection of urinary specimens of asymptomatic patients when not clinically indicated (Givler and Givler, 2019; Nicolle *et al.*, 2019);
- use of specimens as opposed to swabs; swabs pick up insignificant microbes, hold extremely small volumes of the specimen (~0.05 ml), make it difficult

to get bacteria or fungi away from the swab fibres and on to the media, and the inoculum from the swab is often not uniform across several different agar plates (Baron *et al.*, 2013);

* use of swabs for nasopharyngeal and viral respiratory infections (Baron *et al.*, 2013);
* adequate volume to ensure accurate analysis, e.g. blood cultures (Baron *et al.*, 2013);
* correct labelling to ensure reliable interpretation of results (Baron *et al.*, 2013);
* sufficient clinical information and a history of the disease or presenting symptoms (Baron *et al.*, 2013);
* storage at the correct temperature (Baron *et al.*, 2013);
* transport to the laboratory as soon as possible after collection (Baron *et al.*, 2013);
* transport of specimens for virus detection on wet ice and frozen at −80°C if testing is delayed more than 48 hours, although specimens in viral transport media may be transported at room temperature when rapid delivery to the laboratory is guaranteed (Baron *et al.*, 2013);
* collection of blood cultures only when clinically indicated; routine 'surveillance' blood cultures can be expensive and have little clinical value (Kirn and Weinstein, 2013, pp. 513–514);
* avoidance of blood cultures from intravascular devices due to the increased risk of contamination (Kirn and Weinstein, 2013, pp. 513–514).

Case Study 1

A 47-year-old female patient was admitted to hospital with a fractured femur. A urinary catheter was inserted on the day of admission. This catheter was removed on the third day following surgery. On the fifth day following surgery, the patient complained of a burning sensation when passing urine, and suprapubic pain. For two consecutive days, the patient also had a raised temperature of 38°C. A urine sample was sent for microscopy, culture and susceptibility (MCS) and the patient commenced treatment with a Carbapenem.

The preliminary result of the urine sample was:

Gram stain: Gram-negative bacilli, 75,000 leucocytes.
Final laboratory result: Carbapenem-resistant *E. coli* (CPE producer: Oxa-48). Sensitive only to Amikacin and Fosfocymin.

The patient completed 5 days of the Carbapenem as per the original prescription.

Questions:

1. Is this colonization or infection?
2. Do you think the patient needed antibiotics?
3. Was the choice of antibiotics correct? Should the antibiotic have been changed after the sensitivity result became available?

Continued

Case Study 1 Continued.

4. What is the route of transmission of this organism?
5. What type of precautions should be implemented if an infection is indicated?

Answers:

1. Infection – the patient showed signs and symptoms of infection.
2. Yes; the patient showed clinical signs and symptoms of infection.
3. It is recommended that a broad-spectrum antibiotic should be prescribed until the causative organism is known. The treatment can then be de-escalated or changed, based on the susceptibility result. This was not done.
4. Contact (direct and indirect).
5. Contact precautions.

Case Study 2

A 50-year-old patient with diabetes is admitted with a chronic lower-leg ulcer. No clinical signs and symptoms of infection are visible. The healthcare provider removes the bandage, collects a specimen of exudate and commences treatment with intravenous Augmentin (amoxicillin/clavulanic acid) and asks the staff to continue with this treatment for 7 days. The healthcare provider writes the patient's name on the specimen and completes the specimen collection form but omits to add the specimen site or any clinical history. The type of specimen section is completed as 'pus swab'. The healthcare provider, however, forgot to send the specimen to the laboratory. Someone found it the following day and sent it to the laboratory.

After 3 days the results of the culture are returned with the following organisms cultured:

Epithelial cells: +++
S. aureus: Resistant to Methicillin
Pseudomonas aeruginosa: Resistant to Carbapenems and Cephalosporins
Klebsiella pneumoniae: Resistant to Carbapenems

Questions:

1. Was the specimen collected correctly and is it a representative specimen?
2. What additional information is required on the laboratory request form?
3. Was the specimen stored optimally?
4. Was antibiotic treatment indicated?
5. Should the patient have been isolated?

Answers:

1. The wound should first be cleaned with normal saline or sterile water prior to specimen collection to ensure that all debris and pus are removed. A wound swab should then be taken with a zig-zag movement of the wound bed. A wound biopsy is the preferred specimen or, alternatively, a specimen of the wound bed.
Continued

> **Case Study 2** Continued.
>
> **2.** It is important to add the site of the wound (e.g. left lower leg) to the laboratory request form, as well as clinical signs and symptoms, and any antimicrobial treatment that the patient receives.
> **3.** The specimen should be sent to the laboratory as soon as possible after collection. If there is a delay, the wound swab should be refrigerated until sent.
> **4.** When there are no clinical signs and symptoms of infection, antimicrobial treatment is not indicated.
> **5.** The patient should be isolated with contact transmission-based precautions.

Key Points

- The Gram stain guides empiric treatment decisions.
- Antibiotic resistance develops when micro-organisms adapt and grow in the presence of antibiotics and limits the treatment options.
- Surveillance is essential to monitor the prevalence of multidrug-resistant micro-organisms and susceptibility profiles as well as for the identification of populations at risk.
- Vaccination reduces the risk of infection and the need for antimicrobial treatment.
- The correct interpretation of laboratory results is important for optimal treatment and relies on the collection and management of good-quality specimens.

Further Reading

Bauman, R.W. (2017) *Microbiology with Diseases by Body System*, 5th edn. Pearson.

College of Physicians of Philadelphia. The history of vaccines: how vaccines work. Available at: https://www.historyofvaccines.org/content/how-vaccines-work (accessed 13 September 2019).

Core Elements of Hospital Antibiotic Stewardship Programs. Centers for Disease Control and Prevention. Available at: https://www.cdc.gov/antibiotic-use/core-elements/hospital.html (accessed 20 December 2019).

ECDC (2017) *Surveillance of Antimicrobial Resistance in Europe*. Available at: https://ecdc.europa.eu/sites/portal/files/documents/EARS-Net-report-2017-update-jan-2019.pdf (accessed 19 September 2019).

Lewin, R.A and Andersen, R.A. Algae. *Encyclopaedia Britannica*. Available at: https://www.britannica.com/science/algae (accessed 13 September 2019).

Lourdes, P. Norman-McKay (2019) *Microbiology: Basic and Clinical Principles*. Pearson.

Tackling Antimicrobial Resistance 2019–2024: The UK's Five-year National Action Plan. UK Department of Health and Social Care, 2019. Available at: https://www.gov.uk/government/publications/uk-5-year-action-plan-for-antimicrobial-resistance-2019-to-2024 (accessed 20 December 2019).

References

Baron, E.J., Miller, J.M., Weinstein, M.P., Richter, S.S., Gilligan, P.H., *et al*. (2013) A guide to utilization of the microbiology laboratory for diagnosis of infectious diseases: 2013 recommendations by the Infectious Diseases Society of America (IDSA) and the American Society for Microbiology (ASM) (a). *Clinical Infectious Diseases* 57(4), e22–121. DOI: 10.1093/cid/cit278

Bauman, R.W. (2018) *Microbiology with Diseases by Body System*, 5th edn, Pearson.

CDC (Centers for Disease Control and Prevention) (2012) Vaccines: the basics. Available at: https://www.cdc.gov/vaccines/vpd/vpd-vac-basics.html (accessed 13 September 2019).

CDC (2013) Antibiotic resistance threats in the United States, 2013. Available at: https://www.cdc.gov/drugresistance/threat-report-2013/pdf/ar-threats-2013-508.pdf (accessed 13 September 2019).

CDC (2014) Core elements of hospital antibiotic stewardship programs. Available at: http://www.cdc.gov/getsmart/healthcare/implmentation/core-elements (accessed 2 April 2019).

CDC (2016a) List of vaccines used in the United States. Available at: https://www.cdc.gov/vaccines/terms/glossary.html#commimmunity (accessed 13 September 2019).

CDC (2016b) Vaccines by disease. Available at: https://www.cdc.gov/vaccines/vpd/vaccines-diseases.html (accessed 13 September 2019).

CDC (2018a) Introduction to public health surveillance. Available at: https://www.cdc.gov/publichealth101/surveillance.html (accessed 3 July 2019).

CDC (2018b) List of vaccines used in the United States. Available at: https://www.cdc.gov/vaccines/vpd/vaccines-list.html (accessed 3 June 2019).

Chisholm, R.H., Campbell, P.T., Wu, Y., Tong, S.Y.C., McVernon, J. and Geard, N. (2018) Implications of asymptomatic carriers for infectious disease transmission and control. *Royal Society Open Science* 5(2),1–13. Available at: https://royalsocietypublishing.org/doi/pdf/10.1098/rsos.172341 (accessed 13 September 2019).

EC (2017) *A European One Health Action Plan against Antimicrobial Resistance*. Available at: https://ec.europa.eu/health/amr/sites/amr/files/amr_action_plan_2017_en.pdf (accessed 19 September 2019).

ECDC (European Centre for Disease Control and Prevention) (2018) *Long-term Surveillance Strategy 2014–2020*. Available at: https://ecdc.europa.eu/sites/portal/files/documents/LTSS-revised_0.pdf (accessed 19 September 2019).

Forder, A. (2005) Staining reactions. Lecture notes.

Givler, D.N. and Givler, A. (2019) *Asymptomatic bacteriuria*. Available at: https://www.ncbi.nlm.nih.gov/books/NBK441848/ (accessed 19 September 2019).

Holmes, A.H., Moore, L.S.P., Sundsfjord, A., Steinbakk, M.D., Regmi, S. and Karki, A. (2016) Understanding the mechanisms and drivers of antimicrobial resistance. *The Lancet* 387(10014). Available at: http://linkinghub.elsevier.com/retrieve/pii/S0140673615004730 (accessed 19 September 2019).

IACG (Interagency Coordination Group on Antimicrobial Resistance) (2018) Surveillance and monitoring for antimicrobial use and resistance. Available at: https://www.who.int/antimicrobial-resistance/interagency-coordination-group/

IACG_Surveillance_and_Monitoring_for_AMU_and_AMR_110618.pdf?ua=1 (accessed 19 September 2019).

Jansen, K.U. and Anderson, A.S. (2018) The role of vaccines in fighting antimicrobials resistance (AMR). *Human Vaccines & Immunotherapeutics* 14, 2142–2149. DOI: org/10.1080/21645515.2018.1476814

Johnson, A.P. (2015) Surveillance of antibiotic resistance. *Philosophical Transactions of the Royal Society Publishing* 5, 370. Available at: http://dx.doi.org/10.1098/rstb.2014.0080 (accessed 19 September 2019).

Kannan, I. (2016) *Essentials of Microbiology for Nurses*, 1st Edition. Elsevier, India.

Kirn, T.J. and Weinstein, M.P. (2013) Update on blood cultures: how to obtain, process, report and interpret. *Clinical Microbiology and Infection* 19, 513–520. DOI: 10.1111/1469-0691.12180

Kliegman, R.M., St Geme, J.W., Blum, N.J., Shah, S.S., Taskser, R.C., *et al.* (2019) *Nelson Textbook of Pediatrics* (e-book), 21st Edition. Elsevier Health Science, pp. 1347–1366.

Klugman, K.P. and Black, S. (2018) Impact of existing vaccines in reducing antibiotic resistance: primary and secondary effect. *Proceedings of the National Academy of Science of the United States of America* 115(51), 12896–12901. Available at: https://www.pnas.org/content/115/51/12896 (accessed 19 September 2019).

Lipsitch, M. and Siber, G.R. (2016) How can vaccines contribute to solving the antimicrobial resistance problem? *American Society for Microbiology* 3, 1–8. DOI: 10.1128/mBio.00428-16

Lloyd-Price, J., Abu-Ali, G. and Huttenhower, C. (2016) The healthy human microbiome. *Genome Medicine* 8(51), 1–11. Available at: https://genomemedicine.biomedcentral.com/articles/10.1186/s13073-016-0307-y (accessed 2 June 2019).

MacVane, S.H., Hurst, J.M. and Steed, L.L. (2016) The role of antimicrobial stewardship in the clinical microbiology laboratory: stepping up to the plate. *Open Forum Infectious Diseases* 3(4). Available at: https://www.google.com/search?q=The+role+of+antimicrobial+stewardship+in+the+clinical+microbiology+laboratory%3A+stepping+up+to+the+plate#spf=1568922181101 (accessed 19 September 2019).

Mediclinic Southern Africa (2015) *Corporate Policy*. Specimen Collection Guidelines.

Nicolle, L.E., Gupta, K., Bradley, S.F., Colgan, R., DeMuri, G.P. *et al.* Clinical practice guideline for the management of asymptomatic bacteriuria. *Clinical Infectious Diseases* 21 March 2019. Available at: https://www.idsociety.org/practice-guideline/asymptomatic-bacteriuria/ (accessed 2 June 2019).

PHAC (Public Health Agency of Canada) (2013) Routine practices and additional precautions for preventing the transmission of infection in health care. Her Majesty the Queen in Right of Canada, Ottawa. Available at: https://www.canada.ca/en/public-health/services/publications/diseases-conditions/routine-practices-precautions-healthcare-associated-infections.html (accessed 19 December 2019).

Phizer (2014) The value of vaccines in disease prevention. Available at: https://www.pfizer.com/news/hot-topics/the_value_of_vaccines_in_disease_prevention_latin_america_english (accessed 19 September 2019).

Sender, R., Fuchs, S. and Milo, R. (2016) Revised estimates for the number of human and bacteria cells in the body. *PLOS Biology* 14(8). Available at:

https://www.ncbi.nlm.nih.gov/pubmed/27541692 (accessed 19 September 2019).

Singer, A.C., Shaw, H., Rhodes, V. and Hart, A. (2016) Review of antimicrobial resistance in the environment and its relevance to environmental regulators. *Frontiers in Microbiology* 7, 1728. Available at: https://www.ncbi.nlm.nih.gov/pmc/articles/PMC5088501/ (accessed 19 September 2019).

Sood, S. (2013) *Microbiology for Nurses*, 3rd Edition. Elsevier, India.

Tap, J., Mondot, S., Levenez, F., Pelletier, E., Caron, C., *et al.* (2009) Towards the human intestinal microbiota phylogenetic core. *Environmental Microbiology* 11(10), 2574–2584. Available at: https://www.ncbi.nlm.nih.gov/pubmed/19601958 (accessed 19 September 2019).

UKDOH (United Kingdom Dept of Health) (2014) *UK 5-year Antimicrobial Resistance (AMR) Strategy 2013–2018: Measuring Success*. Available at: https://assets.publishing.service.gov.uk/government/uploads/system/uploads/attachment_data/file/662189/UK_AMR_3rd_annual_report.pdf (accessed 19 September 2019).

WHO (World Health Organization) (2015) *Global Action Plan on Antimicrobial Resistance*. Available at: https://apps.who.int/iris/bitstream/handle/10665/193736/9789241509763_eng.pdf?sequence=1 (accessed 19 September 2019).

WHO (2016) Why is vaccination important for addressing antibiotic resistance? Available at: https://www.who.int/features/qa/vaccination-antibiotic-resistance/en/ (accessed 19 September 2019).

WHO (2018) *Global Antimicrobial Resistance Surveillance System (GLASS) Report*. Available at: https://apps.who.int/iris/bitstream/handle/10665/279656/9789241515061-eng.pdf?ua=1 (accessed 19 September 2019).

Wright, G.D. (2007) The antibiotic resistome: the nexus of chemical and genetic diversity. *Nature Reviews Microbiology* 5(3), 175–186. DOI: 10.1038/nrmicro1614

Antimicrobials and Antimicrobial Resistance

3

Maria Clara Padoveze[1],*, Ligia Maria Abraão[2] and Rosely Moralez de Figueiredo[3]

[1]Associate Professor, Department of Collective Health Nursing, University of São Paulo, Brazil; [2]Coordinator, Infection Control Service (Corporate Team), Americas Medical Services, São Paulo, Brazil; [3]Professor, Federal University of São Carlos, Brazil

Objective: For the student to understand the core knowledge underpinning the concept of antimicrobial resistance and apply this knowledge to nursing practice to help prevent it.

Introduction

Despite the benefits of antibiotic use in healthcare, including reduced mortality, increased life expectancy and as adjuvants to other therapies, such as chemotherapy, infections caused by multidrug-resistant bacteria are now more frequent, globally, with therapeutic options to treat them increasingly limited (Blair *et al.*, 2015). Evidence points out that resistance to antimicrobials is part of the natural evolution of bacteria, and is likely to have existed before their discovery by humans (Fair and Tor, 2014). However, the contribution of humans to this evolutionary process and selective pressure is undoubtedly significant, which will be the core approach of this chapter. Antimicrobial resistance (AMR) is the capability of a micro-organism to overcome the action of an antimicrobial. When micro-organisms are exposed to antimicrobials, the susceptible strains will be killed, leading to the survival of the resistant ones. AMR is driven by the overuse of antibiotics in medicine, livestock farming and agriculture, and poultry production. Moreover, a further contributory factor is the lack of development of new antibiotics by the pharmaceutical industry (Blair *et al.*, 2015; Pontes *et al.*, 2018).

*Corresponding author: padoveze@usp.br

Nurses have a crucial role to play in the early recognition of infection and actively take part in antimicrobial therapy (Courtenay *et al.*, 2018, 2019). It is therefore essential that they understand the underlying concepts of antimicrobial therapy and the process that leads to AMR.

Signs and Symptoms of Infection

An infection results from the imbalance between the mechanisms employed by the micro-organisms causing diseases and the host response to prevent this aggression. The outcome of this episode will depend on the number of micro-organisms and their ability to find the mode of entry, overcome host defences, invade tissue and produce toxins. Micro-organisms can be installed in a susceptible host by a number of routes including inhalation, ingestion, direct contact, direct inoculation and rupture of skin barriers (Fischbach and Dunning, 2015).

During healthcare, numerous situations allow pathogens to get in contact with a host. Hence, understanding the micro-organisms' routes of transmission, the pathogenic mechanisms and the susceptibility of the host is key to guiding the prevention measures adopted during care. The presence of pathogenic micro-organisms in a patient does not necessarily characterize an infection. If signs and symptoms of infection are absent, the patient can be considered colonized by pathogens. Thus, colonization is the process by which micro-organisms are present in the host, with a certain level of multiplication, but without producing clinical disease. Infection, however, involves the invasion of micro-organisms in the host's body tissues, with the generation of an inflammatory and immunological response, leading to clinical disease and resulting in signs and symptoms such as fever, purulent exudate from a wound, high blood cell count or pneumonia (Dani, 2014).

Not all colonization, however, results in infection. Some people may become temporarily or permanently colonized with pathogenic micro-organisms but never develop symptomatic disease, whereas others can become seriously ill and may even die due to the infection. The process of colonization is complex and sometimes involves changes in the colonizing strains, only detected by molecular studies (Padoveze *et al.*, 2008). In many situations, individuals immediately progress from colonization (or after a period of colonization) to symptomatic disease. The evolution of an initial colonization is frequently difficult to predict. Immune status at the time of exposure to an infectious agent and aspects of the pathogen's own virulence would ultimately define the evolution of the case (Siegel *et al.*, 2007).

Healthcare professionals should be aware of the early signs and symptoms of an infectious process (Singer *et al.*, 2016). The clinical presentation may be restricted to a certain anatomical location and include local signs and symptoms such as pain, erythema (redness), presence of purulent exudate (pus) and abscess formation. General signs and symptoms such as fever, tachycardia, sweating and changes in biomarkers reflected in laboratory tests (such as haemogram and cerebrospinal fluid) may indicate a systemic infection. The signs and symptoms of an infection vary depending on which area of the body is infected. Table 3.1

Table 3.1. Common infection sites and main signs and symptoms.

Major infection sites	Main signs and symptoms
Bloodstream	Fever, tachycardia, tachypnoea, hypotension, mental confusion, oliguria
Digestive tract	Nausea, vomiting, diarrhoea, abdominal pain
Respiratory tract	Tachypnoea, presence or change of secretion (changes in amount, colour, appearance)
Skin	Erythema, heat, pain, pus, macules, papules, vesicles
Urinary tract	Dysuria, polyuria, low abdominal pain
Wound	Erythema, heat, pain, pus, necrosis

describes common anatomical sites of infection and their main signs and symptoms. Of note, infections in neonates and children may exhibit atypical signs and symptoms, which may result in a diagnostic delay.

Nurses have a pivotal role to play in the early identification of infection. In in-patient settings, nurses provide bedside care around the clock, and are therefore in a privileged position to detect any changes, albeit slight, in the patient's condition. In primary and community care settings, nurses are usually the first and often the only point of contact for patients and families with the healthcare system.

Several nursing care activities, other than the measurement of vital signs and physical examination, provide opportunities to facilitate early detection of infection. For example, when administering intravenous medication, signs of inflammation may be observed at the catheter insertion site. If dressing a wound, the presence of purulent secretion or necrosis may be evidenced. Erythema, heat, or macules and papules may be noted when bathing a patient. During the emptying of a catheter bag, nurses can perceive changes in the urine, including pyuria.

All healthcare professionals should recognize the importance of adequate specimen collection during relevant stages of antimicrobial use (prior, during and post-antibiotic treatment). The optimal time for specimen collection varies according to the type of infection; however, for diagnostic purposes, sample collection is more profitable in the early stages of an infection episode (O'Donnell and Guarascio, 2017). Nurses should be aware of the need to correlate the signs and symptoms of infection and the optimal time for specimen collection. Ideally, nurses should ensure that cultures are performed before starting antibiotics, although antibiotic administration should not be delayed if this would affect the clinical outcome for the patient. Please refer to Pollack and Srinivasan (2014).

The adequate conservation of samples of blood, urine, faeces and secretions, and their delivery to the laboratory in a timely manner, is essential to ensure the quality of the samples collected and the results obtained. Negligence or carelessness in sample collection can lead to misdiagnosis, wasting laboratory time and delayed effective treatment (Fischbach and Dunning, 2015). This aspect is detailed in Chapter 2.

Qualified, sensitized and attentive nursing staff, with an appropriate workload, have demonstrated adherence to good practices with a decrease in the risk of healthcare-associated infection (Siegel *et al.*, 2007).

How Antimicrobials Work

Generally speaking, antimicrobials include any agent (such as antibiotics, antivirals, antifungals and antimalarials) with biological activity against micro-organisms. Antibiotic is a type of antimicrobial substance active against bacteria. Antimicrobials may be produced biologically by micro-organisms or synthetically by chemists. During the stages of pharmaceutical development, the antimicrobial activity of the drug is usually tested against several groups of micro-organisms to determine its spectrum of activity. A broad-spectrum drug is one that has activity against Gram-positive as well as Gram-negative species. A narrow-spectrum drug has activity against only one group of micro-organisms or only one species. The antimicrobial agents may either kill micro-organisms (-cidal) or inhibit their growth (-static), e.g. bactericidal or bacteriostatic.

The biochemical activities of micro-organisms are so similar to those in mammalian cells that the drug will affect both cell types. However, most antimicrobials have selective toxicity, i.e. show greater affinity to the microbial component than to the mammalian cells. The therapeutic index refers to the level of selective toxicity of a given drug comparing the blood concentration at which a drug causes a therapeutic effect to the amount of toxicity produced in humans (Tamargo *et al.*, 2015). A low therapeutic index indicates that the concentration of the drug that is therapeutic is also harmful to host tissue.

A complete description of the morphology and physiology of the micro-organisms' cells is beyond the scope of this chapter (see Chapter 2 instead). However, it is important to point out that viruses, for instance, are primarily intracellular pathogens, while bacteria may have a free life. This means that viruses need to use the structures and metabolism of host cells in order to survive and reproduce. Bacteria, on the other hand, can survive and reproduce without being sheltered in a host cell, which ensures their ability to survive in the environment. Among the relevant differences between Gram-positive and Gram-negative bacteria, the latter have thinner cell walls that are surrounded by lipid membrane, while Gram-positive bacteria have a tough and rigid mesh cell wall that surrounds the cytoplasmic membrane (Kapoor *et al.*, 2017). In bacterial cells, protein synthesis occurs in ribosomes from a messenger Ribonucleic acid (mRNA) in a process called translation. The ribosomes are a cell structure that makes protein. Proteins are needed for several cell functions such as repairing damage or directing chemical process. Bacterial ribosome represents one of the major targets for antibiotics in the cell. This structure is composed of three Ribonucleic acid (RNA) chains (16S, 23S and 5S) and more than 50 proteins assembled in two individual subunits, the small one known as 30S and the large one as 50S (Lin *et al.*, 2018).

Classification of antimicrobials may vary according to the targeted micro-organisms (i.e. antibiotics, antivirals, antifungals, antimalarials), their molecular structure or their antimicrobial mechanisms (i.e. their ability to affect various essential cellular functions) (Hoerr *et al.*, 2016). Table 3.2 provides some examples of the mechanism of action of currently available antimicrobials.

- *Inhibitors of cell wall synthesis:* Antimicrobials may target the bacterial cell wall, mainly interfering with the building of the peptidoglycan chain. The penicillin-binding protein (PBP) is a transpeptidase, which plays a role in the cell wall building process. However, the β-lactam ring, that is part of the core structure of several antibiotic families, interacts with PBP, which then is unavailable for the synthesis of new peptidoglycan. Lysis of bacteria occurs due to disruption of the peptidoglycan layer (Dowling *et al.*, 2017).

- *Inhibitors of cytoplasmic membrane function:* Antimicrobials with affinity for lipids can bind irreversibly to cytoplasmic membrane sterols, such as ergosterol; others interfere with lipid biosynthesis. These actions lead to alterations in permeability, resulting in loss of nutrients and other compounds.

- *Inhibitors of protein synthesis:* The process of inhibition of protein synthesis can occur through a variety of ways of binding to ribosomes, which causes a misreading of mRNA, resulting in an abnormal protein or impairment of protein synthesis. Once incorporated in the cytoplasmic membrane, some of the abnormal protein will produce pores that enable more antibiotic to enter the bacterial cell. As more antibiotic reaches the cytoplasm and binds to ribosomes, the process continues up to shutting down the cell. Antibiotics such as macrolides, aminoglycosides and tetracyclines bind to the 30S subunit of ribosome, whereas chloramphenicol binds to the 50S subunit. In these binds, the aminoacyl and peptidyl transfer may be blocked. The binding of antimicrobial to free ribosomes allows the formation of a small peptide but prevents further elongation of the RNA chain (Dowling *et al.*, 2017; Kapoor *et al.*, 2017).

- *Inhibitors of nucleic acid synthesis:* Some antibiotics exert their antibacterial effects by disrupting Deoxyribonucleic acid (DNA) synthesis and causing lethal double-strand DNA breaks during the replication process. Antimicrobials interfere with nucleic acid synthesis by using several mechanisms:
  ○ binding to a subunit of the DNA-dependent RNA polymerase and therefore interfering with the initiation of transcription of this enzyme;
  ○ inhibition of enzymes such as Topoisomerase II (DNA gyrase) and Topoisomerase IV, so impairing DNA replication (Kapoor *et al.*, 2017);
  ○ phosphorylation of 5-fluorouracil and incorporation into RNA, interfering with normal protein synthesis;
  ○ inhibition of the viral primary transcription process, and preventing uncoating of the virus capsid;
  ○ inhibition of early viral replication step and subsequent impairment of viral nucleic acid synthesis;
  ○ acting as analogue of thymidine to be incorporated into viral DNA;
  ○ inhibition of enzymes that convert viral RNA into viral DNA;
  ○ blocking the pathway for folic acid synthesis, which inhibits DNA synthesis (Tenover, 2006).

Table 3.2. Mechanisms of antimicrobial action according to class compound and respective antimicrobial drug.

Mechanism of antimicrobial action	Class compound	Antimicrobial drug
Inhibitors of cell wall synthesis	Beta-lactams	Penicillins, ampicillin, amoxicillin, carbenicillin, methicillin
	Beta-lactams	Cephalosporins (cephalothin, cefamandole, cefotaxime, cefoxitin, cefazolin, cefoperazone, cefixime, cefprozil, cefpodoxime, ceftaroline, ceftolozane-tazobactam, ceftazidime-avibactam)
	Beta-lactams	Carbapenems (imipenem, meropenem, ertapenem, doripenem)
	Glycopeptides	Vancomycin, teicoplanin
Inhibitors of cytoplasmic membrane function	Polyenes	Amphotericin B, nystatin, candicidin, pimaricin, trichomycin, hamycin*
	Azoles	Ketoconazole, fluconazole, itraconazole*
	Polymixins	Polymixin B, polymixin E (colistin)
Inhibitors of protein synthesis	Aminoglycosides	Streptomycin, kanamycin, gentamycin, tobramycin, sisomycin, amikacin, neomycin, fortimicin A, netilmicin, 5-episisomicin, spectinomycin
	Tetracyclines	Tetracycline, doxycycline
	Amphenicols	Chloramphenicol
	Macrolides	Erythromycin, azithromycin, spiramycin, josamycin, roxithromycin, clarithromycin
	Lincosamides	Lincomycin, clindamycin
	Oxazolidinones	Linezolid
Inhibitors of nucleic acid synthesis	Ansamycins	Rifampicin, rifamycin B
	Quinolones	Nalidixic acid, fluoroquinolones (norfloxacin, ciprofloxacin, ofloxacin)
	Pyrimidines	Flucytosine*
	Amines	Amantadine, rimantadine**
	Virazole	Ribavirin**
	Nucleoside analogues	Idoxuridine, thymidine, vidarabine, acyclovir, zidovudine (azidothymidine or AZT)**
Antimetabolites	Sulfonamides	Sulfanilamide, sulfadiazine, sulfamethoxazole, sulfathiazole, sulfasoxazole, sulfapyridine

Continued

Table 3.2. Continued.

Mechanism of antimicrobial action	Class compound	Antimicrobial drug
	Sulfones	Dapsone
	Aminopyrimidines	Trimethoprim, ormetoprim
	Nitrofurans	Nitrofurantoin
	Others	Ethambutol, isoniazid (INH) Para-aminosalicyclic acid

*antifungals; **antivirals.

- *Antimetabolites:* These are compounds structurally similar to normal cellular metabolites and can compete with them for attachment to enzymes (Hoerr *et al.*, 2016; Kapoor *et al.*, 2017). For example:
  - similarity with para-aminobenzoic acid (PABA), competing for the enzyme dihydrofolate synthase and therefore preventing the synthesis of folic acid (Dowling *et al.*, 2017);
  - structural analogue of pteridine, preventing the synthesis of folic acid, and interfering with amino acid, and purine and pyrimidine synthesis;
  - structural analogue of nicotinamide, impairing the synthesis of nicotinamide adenine dinucleotide;
  - binding and inhibition of β-lactamase, resulting in a restoration of the antimicrobial activity (Dowling *et al.*, 2017).

Appropriate Antimicrobial Use

As previously mentioned, antibiotic resistance is the capacity of bacteria to survive and replicate in the presence of antibiotics that normally act to inhibit or kill them (Pontes *et al.*, 2018). There are two types of antimicrobial resistance: intrinsic and acquired (Dowling *et al.*, 2017).

- *Intrinsic resistance* is related to the innate ability of all or almost all prokaryotes (unicellular organisms that lack organelles or internal membrane-bound structures) to resist specific drugs; i.e. it occurs naturally in bacterial genomes.
- *Acquired resistance* may arise from spontaneous chromosomal mutations or exchange of genetic elements among micro-organisms. The latter situation occurs through mobile genetic elements, such as plasmids, integrons, transposons or genomic islands. Both these mechanisms of resistance are important for the dissemination of resistance between different species. They also contribute to the bacterial genome evolution. Acquired resistance is also associated with a gradual increase in antibiotic concentrations (Pontes *et al.*, 2018). This gradual increase creates an environmentally selective pressure and bacteria develop reversible drug-resistance profiles. It is important to note that, in small antibiotic doses (non-lethal concentrations), bacteria

can survive and grow normally. Nevertheless, non-lethal concentrations can induce specific resistance mechanisms (Lin *et al.*, 2015; Pontes *et al.*, 2018). These mechanisms are the dependant of the type of drug used and may induce resistant mechanisms against both the drug in use and related drugs (Pontes *et al.*, 2018). This phenomenon points out an implication for nursing practice highlighting the importance of proper antibiotic administration. Missing doses can lead to suboptimal drug concentration in the target-body sites, inducing the expression of resistant mechanisms or favouring the selection of resistant strains.

The main mechanisms found in bacteria to impair the action of antimicrobials are: (i) reduction of bacterial membrane permeability; (ii) increases in both expression and activity of efflux pump systems; (iii) synthesis of enzymes that are able to destroy or modify the drug; (iv) modification, substitution or disruption of antibiotic bacterial targets; and (v) biofilm formation.

Mechanisms (i)–(iv) are associated with antimicrobial resistance to the main antibiotic groups used in clinical practice: β-lactams, glycopeptides and aminoglycosides (Lin *et al.*, 2015; Pontes *et al.*, 2018). These mechanisms are discussed below.

- *Reduction of bacterial membrane permeability:* Cellular membrane is an important compound in a micro-organism's structure. This membrane has an intrinsic permeability that allows nutrient intake and acts as a first barrier to external agents. When the permeability on the cell surface is low, antibiotic blocking occurs, so drugs cannot achieve their targets. Compared to Gram-positive species, Gram-negative bacteria are usually less permeable to many antibiotics, as their outer membrane forms a permeability barrier. For instance, the glycopeptide antibiotic vancomycin inhibits peptidoglycan cross-linking. It does not occur in Gram-negative bacteria, in which this drug cannot cross the membrane and reach these substances in the periplasm (Blair *et al.*, 2015; Pontes *et al.*, 2018).

- *Synthesis of enzymes that are able to destroy or modify the drug:* A large number of enzymes that have been identified can degrade and modify antibiotics of different classes, such as β-lactams, aminoglycosides, amphenicols and macrolides. There are also subclasses of enzymes that can destroy or degrade different antibiotics which belong to the same class. An example is the β-lactam antibiotics group, where penicillin, cephalosporins, clavams, carbapenems and monobactams are hydrolysed by a variable range of β-lactamases, which are enzymes that destroy the β-lactams (Blair *et al.*, 2015; Lin *et al.*, 2015; Pontes *et al.*, 2018).

- *Modification, substitution or disruption of antibiotic bacterial targets:* This is one of the most common mechanisms of antimicrobial resistance. Most antibiotics bind to the bacterial targets in a specific way, thus preventing their normal role. Modifications in these targets, characterized by small mutations that prevent efficient antibiotic binding, can confer resistance (Blair *et al.*, 2015; Pontes *et al.*, 2018).

- *Increases in both expression and activity of efflux pump systems:* Bacterial efflux pumps are an important mechanism of resistance in which many antibiotics are actively transported out of the cells to the environment prior to reaching their intended targets. Efflux pumps are a kind of protein that constitutes all bacterial plasma membranes. When an over-expression (abnormal increase in the expression of a certain gene) occurs, the efflux pumps can provide a high level of resistance to antibiotics that were previously useful in the clinical practice. This mechanism has been seen since 1990 among some bacteria including *Enterobacteriaceae, Pseudomonas aeruginosa* and *Staphylococcus aureus* (Blair *et al.,* 2015; Lin *et al.,* 2015; Pontes *et al.,* 2018).

The speed with which micro-organisms have acquired resistance by increasingly diverse mechanisms demonstrates that better practices related to antibiotic use within the multidisciplinary health team are necessary (Pontes *et al.,* 2018).

By recognizing the mechanisms that lead to resistance, and the selective pressure exerted on the environment by the excessive use of antimicrobials, nurses can adopt a proactive attitude regarding antibiotic therapy. As part of their activities, nurses should obtain an allergy history from the patient to support optimal prescribing of antimicrobials, as well as monitoring and reporting adverse events of antimicrobial therapy. Nurses should review daily the clinical condition of patients, perform a proper choice of vascular access according to the therapy, and actively take part in multidisciplinary team discussions regarding antibiotic treatment, indication and duration. Please refer to Pollack and Srinivasan (2014).

The group of micro-organisms that each antimicrobial can target is identified during the pharmaceutical development. However, there are various behaviours of susceptibility and resistance among micro-organisms. In clinical practice, the microbiological examination of specimens collected from infected or colonized patients would provide information about the species that are the probable etiologic agent, as well as the antibiogram or antibiotic susceptibility report. Due to their characteristics of intracellular reproduction, culture tests and antibiograms are not routinely requested for viruses in clinical settings, but may be performed in research centres or public health laboratories.

The antibiogram will guide the choice of drug therapy by providing specific information regarding the susceptibility of a particular pathogen to certain antibiotics, including the minimum inhibitory concentration (MIC). MICs have established breakpoints to define susceptibility or resistance according to international organizations such as the Clinical and Laboratory Standards Institute (CLSI) and European Committee on Antimicrobial Susceptibility Testing (EUCAST) (O'Donnell and Guarascio, 2017). A timely microbiological examination with antibiogram will favour informed decisions earlier in the course of the infection and positively influence treatment outcomes. In many cases, however, antibiotics are prescribed and administered empirically, as microbiologic results may not be available at the time of making the clinical decision. Once the antibiogram is available, if etiologic agents are susceptible to antibiotic therapy, it would be possible to adjust the treatment accordingly, i.e. antibiotic de-escalation.

Aiming to support the prudent use of antimicrobials in both human and veterinary medicine, the World Health Organization (WHO) has published a list of critically important antimicrobials for human medicine (WHO, 2017). Nurses should be aware of the antimicrobials that have restricted indication and the need to save them for specific purposes. For instance, isoniazid is a drug used solely to treat tuberculosis or other mycobacterial diseases. Another example is rifampicin, which is considered as limited therapy as part of the treatment of mycobacterial diseases, including tuberculosis. Both these drugs are categorized as critically important antimicrobials (WHO, 2017). One strategy to combat the increasing antimicrobial resistance is the discovery of new antimicrobials as well as finding strategies to expand the useful life of those in existence. Bacteria, however, possess a great diversity of genes that allow them, at any time, to counteract the action of newly formulated or discovered antibiotic compounds (Lin *et al.*, 2015).

Finally, there is a significant worldwide effort towards a 'One Health' approach to defeat antimicrobial resistance. This approach refers to designing and implementing programmes, policies, legislation and research in which multiple sectors (e.g. human and animal health) work together to achieve better public outcomes (Ryu *et al.*, 2017). Nurses are the biggest healthcare workforce in the world; they are in a unique position to embrace the concept of One Health and be the propellant for a more collaborative, multidisciplinary and global effort against antimicrobial resistance (Premji and Hatfield, 2016).

Conclusion

Nurses are frontline professionals engaged in the fight against antimicrobial resistance. This includes active participation in early detection of infection, and taking responsibility regarding antimicrobial treatment to optimize the use of antimicrobial agents.

Case Study 1: Antimicrobial Stewardship in Action

A three-year-old child arrived at the Primary Health Care Unit (PHCU) with clear/ transparent rhinorrhoea and fever (38°C). The mother stated that the child had been tearful and had had a poor appetite for 3 days. After a medical evaluation that revealed regular overall clinical condition and hydrated mucous membranes with no signs of bacterial infection in the ears, throat and lungs, the child was discharged from the PHCU. The medical prescription included nasal saline, antipyretic (to reduce fever discomfort) and the recommendation to reinforce oral hydration and to return within 2 days for reassessment. The mother expressed concern about the sufficiency of the treatment. She did not agree with the prescription and questioned why antibiotics were not prescribed to prevent worsening of the child's condition.

Continued

Case Study 1: Continued.

Questions and answers

What were the clinical characteristics that pointed out that there was no bacterial infection?

Answer: There was a clear/transparent rhinorrhoea. Medical evaluation revealed regular overall clinical condition and hydrated mucous membranes with no signs of bacterial infection in the ears, throat and lungs.

What arguments would you use to reassure the mother that there is no need for antibiotics?

Answer: The clinical presentation suggested a viral upper airways infection, which is usually self-limiting, with spontaneous remission within a few days. Antibiotics do not act on viruses, and therefore would not improve the child's condition, and may even worsen appetite. In addition, the inappropriate use of antibiotics could contribute to select antimicrobial resistant strains, decreasing the treatment options for eventual future bacterial infections. Emphasize the need for the child's reassessment within 2 days or earlier in the case of any worsening in the child's overall clinical condition.

Case Study 2

An obese patient with diabetes arrived at the emergency room with pulmonary thromboembolism. The patient had no other signs. Intubation and mechanical ventilation were carried out. On the fifth day of hospitalization, the patient developed a fever with a large amount of purulent secretion on tracheal aspirations. Samples of blood and tracheal secretions (semi-quantitative method) were collected for culture. The physician prescribed vancomycin and meropenem empirically. The results of the microbiological exams revealed *Klebsiella pneumoniae* sensitive to the antibiotics tested.

Questions and answers

What are the characteristics that point to an ongoing infection?

Answer: The patient presented with fever and a change in respiratory secretions and had an invasive device that would reduce the body's natural immunity. Because clinical signs and symptoms were not present on admission, and as the length of hospital stay was greater than 72 hours, this was considered a healthcare-associated infection.

What argument would you use to discuss with the physician the need to review the antimicrobial prescription?

Answer: Despite being a healthcare-associated infection, the micro-organism identified is Gram-negative and susceptible to other drugs. De-escalation of antibiotics and exclusion of vancomycin (which is mainly addressed to Gram-positive) should be considered.

Key Points

- An infection results from the imbalance between the mechanisms employed by the micro-organisms causing diseases and the host response to prevent this aggression.
- Infections may be detected by specific local or systemic signs and symptoms.
- Antimicrobials' mode of action includes inhibition of cell wall synthesis, inhibition of cytoplasmic membrane function, inhibition of protein synthesis, inhibition of nucleic acid synthesis and antimetabolite activity.
- Antimicrobial resistance may be intrinsic or acquired.
- The main mechanisms found in bacteria to impair the action of antimicrobials are: reduction of bacterial membrane permeability, increases in both expression and activity of efflux pump systems, synthesis of enzymes which are able to destroy or modify the drug; modification, substitution or disruption of antibiotic bacterial targets, and biofilm formation.

Further Reading

Casadevall, A. and Pirofski, L.A. (2000) Host-pathogen interactions: basic concepts of microbial commensalism, colonization, infection, and disease. *Infection and Immunity* 68(12), 6511–6518.

WHO (World Health Organization). One health. Available at: https://www.who.int/features/qa/one-health/en (accessed 14 September 2019).

WHO (2018) *WHO Competency Framework for Health Workers' Education and Training on Antimicrobial Resistance*. Available at: http://apps.who.int/medicinedocs/documents/s23443en/s23443en.pdf (accessed 14 September 2019).

References

Blair, J.M., Webber, M.A., Baylay, A.J., Ogbolu, D.O. and Piddock, L.J. (2015) Molecular mechanisms of antibiotic resistance. *National Review of Microbiology* 13(1), 42–51.

CDC (Centers for Disease Control and Prevention) (2014) Core elements of hospital antibiotic stewardship programs. Atlanta, Georgia: US Department of Health and Human Services, CDC [cited 02/04/2019]. Available at: http://www.cdc.gov/getsmart/healthcare/implementation/core-elements (accessed 10 September 2019).

Courtenay, M., Lim, R., Castro-Sánchez, E., Deslandes, R., Hodson, K., *et al.* (2018) Development of consensus-based national antimicrobial stewardship competencies for UK undergraduate healthcare professional education. *Journal of Hospital Infection* 100(3), 245–256.

Courtenay, M., Castro-Sánchez, E., Gallagher, R., McEwen, J., Bulabula, A.N.H., *et al.* (2019) Development of consensus based international antimicrobial stewardship competencies for undergraduate nurse education. *Journal of Hospital Infection*, 103(3): 244–250.

Dani, A. (2014) Colonization and infection. *Central European Journal of Urology* 67(1), 86–87.

Dowling, A.M., Adley, C. and O'Dwyer, J. (2017) Antibiotics: mode of action and mechanisms of resistance. In: Méndez-Filas, A. (ed.) *Antimicrobial Research: Novel Bioknowledge and Educational Programs.* Formatex Research Center, Badajoz, Spain.

Fair, R.J. and Tor, Y. (2014) Antibiotics and bacterial resistance in the 21st century. *Perspectives on Medicinal Chemistry* 28(6), 25–64.

Fischbach, F. and Dunning, M. (2015) *Nurse's Quick Reference to Common Laboratory and Diagnostic Tests,* 6th Edition. Wolters Kluwer Health/ Lippincott Williams & Wilkins, Philadelphia, Pennsylvania.

Hoerr, V., Duggan, G.E., Zbytnuik, L., Poon, K.K., Grosse, C., *et al.* (2016) Characterization and prediction of the mechanism of action of antibiotics through NMR metabolomics. *BMC Microbiology* 16, 82.

Kapoor, G., Saigal, S. and Elongavan, A. (2017) Action and resistance mechanisms of antibiotics: a guide for clinicians. *Journal of Anaesthesiology Clinical Pharmacology* 33(3), 300–305.

Lin, J., Nishino, K., Roberts, M.C., Tolmasky, M., Aminov, R.I. and Zhang, L. (2015) Mechanisms of antibiotic resistance. *Frontiers in Microbiology* 6, 34.

Lin, J., Zhou, D., Steitz A.T., Polikanov, Y.S. and Gagnon, M.G. (2018) Ribosome-targeting antibiotics: modes of action, mechanisms of resistance, and implications for drug design. *Annual Review of Biochemistry* 87(1).

O'Donnell, L.A. and Guarascio, A.J. (2017) The intersection of antimicrobial stewardship and microbiology: educating the next generation of health care professionals. *FEMS Microbiology Letters* 364(1).

Padoveze, M.C., de Jesus Pedro, R., Blum-Menezes, D., Bratfich, O.J. and Moretti, M.L. (2008) Staphylococcus aureus nasal colonization in HIV outpatients: persistent or transient? *American Journal of Infection Control* 36(3), 187–191.

Pollack, L.A. and Srinivasan, A. (2014) Core elements of hospital antibiotic stewardship programs from the Centers for Disease Prevention and Control. *Clinical Infectious Diseases* 59(3), S97–S100. Available at: https://academic. oup.com/cid/article/59/suppl_3/S97/318001 (accessed 14 September 2019).

Pontes, D.S., de Araujo, R.S.A., Dantas, N., Scotti, L., Scotti, M.T., de Moura, R.O. and Mendonca-Junior, F.J.B. (2018) Genetic mechanisms of antibiotic resistance and the role of antibiotic adjuvants. *Current Topics in Medicinal Chemistry* 18(1), 42–74.

Premji, S.S. and Hatfield, J. (2016) Call to action for nurses/nursing. *BioMed Research International* 2016, art. ID 3127543.

Ryu, S., Kim, B.I., Lim, J.S., Tan, C.S. and Chun, B.C. (2017) One health perspective on emerging public health threats. *Journal of Preventive Medicine and Public Health* 50(6), 411–414.

Siegel, J., Rhinehart, E., Jackson, M., Chiarello, L. and HICPAC (2007) *Healthcare Infection Control Practices Advisory Committee: 2007 Guidelines for Isolation Precautions: Preventing Transmission of Infectious Agents in Healthcare Settings.* Available at: https://www.cdc.gov/niosh/docket/archive/pdfs/NIOSH-219/0219-010107-siegel.pdf (accessed 15 September 2019).

Singer, M., Deutschman, C.S., Seymour, C.W., Shankar-Hari, M., Annane, D., *et al.* (2016) The third international consensus definitions for sepsis and septic shock (Sepsis-3). *Jama* 315(8), 801–810.

Tamargo, J., Le Heyzey, J.Y. and Mabo, P. (2015) Narrow therapeutic index drugs: a clinical pharmacological consideration to flecainide. *European Journal of Clinical Pharmacology* 71, 549–567.

Tenover, F. (2006) Mechanisms of antimicrobial resistance in bacteria. *American Journal of Infection Control* 34(5), S3–S10.

WHO (World Health Organization) (2017) *WHO Advisory Group on Integrated Surveillance of Antimicrobial Resistance. Critically Important Antimicrobials for Human Medicine*, 5th Edition. World Health Organization, Geneva, Switzerland.

The Diagnosis of Infection and the Use of Antibiotics

4

Jo McEwen[1],*, Heather Kennedy[2] and Nykoma Hamilton[3]

[1]Advanced Nurse Practitioner, Antimicrobial Stewardship, NHS Tayside, Dundee, UK; [2]Advanced Antimicrobial Pharmacist, NHS Tayside, Dundee, UK; [3]Infection Prevention and Control Nurse, NHS Fife, Kirkcaldy, UK

Objective: For the student to demonstrate knowledge of how infections are diagnosed and the appropriate use of antimicrobials, and to apply this knowledge to nursing practice to support the accurate diagnosis of infection and the appropriate use of antimicrobials.

Microbiology Samples

Appropriate biological sampling is an integral part of nursing care and a key contribution of nurses to antimicrobial stewardship (AMS). The diagnosis of infection is aided by results obtained from patient samples, which can identify the pathogen responsible and often its antimicrobial susceptibility pattern. Such identification allows appropriate antimicrobial treatment to be identified, enabling targeted and individualized therapy including de-escalation to narrower spectrum antimicrobials, switch from intravenous to oral therapies, or treatment cessation altogether.

Nurses are routinely responsible for collecting information from physiological observations including temperature, blood pressure, pulse, oxygen saturation, respiratory rate and urine output. By using these results in conjunction with an assessment of the patient for any signs and symptoms of infection, nurses can quickly detect any deterioration in the patient's condition and the need for further biological sampling. Ideally, samples should be obtained prior to starting antimicrobial therapy; however, therapy should not be delayed, as doing so may be detrimental to the patient.

*Corresponding author: jomcewen@nhs.net

Samples are a useful diagnostic tool, but their use must be justified within the context of the clinical presentation of the patient and not employed routinely. Inappropriate sampling can have financial implications as well as exposing the patient to unnecessary interventions that could provide inaccurate results. Nurses can aid AMS by having an awareness of such benefits and disadvantages, conducting a person-centred assessment and ensuring that samples are appropriate and contribute towards further diagnosis and therapy (Inglis, 2003).

Sample collection plays an important role in diagnosis and management of infections. However, sample contamination can have an adverse impact upon the result and can lead to inappropriate antimicrobial treatment. Contamination of blood culture bottles is a common problem within healthcare, with some studies reporting contamination rates of up to 50% (Raja *et al.*, 2009). National Health Service Education for Scotland (NES) estimates that within Scotland the rate of contamination is 10%, whilst optimal rates should be less than 3% for samples in hospitalized patients (NHS Education for Scotland, 2017). One of the most frequent sources of contamination is the skin of the patient (Bentley *et al.*, 2016). Bacteria that normally live on the skin or are from the hands of staff taking the sample may enter the blood as a result of inappropriate skin cleansing prior to blood sampling. Healthcare staff should always follow standard infection control precautions (SICPs) and use transmission-based precautions (TBPs) where required. This includes (but is not limited to) good hand hygiene and use of personal protective equipment (PPE), such as gloves and aprons.

Common samples obtained by nursing staff include:

- blood cultures;
- urine samples;
- stool samples;
- respiratory samples – either as viral throat swab or sputum; and
- wound swab or wound biopsy.

Obtaining any sample requires nurses to use SICPs and also understand which container is appropriate for the sample being collected. Using the incorrect sample container could inadvertently lead to a delay in diagnosis and appropriate antimicrobial therapy. Local sampling guidelines should be followed. Nursing staff should communicate to patients why a sample is required and obtain informed consent from them.

Urinalysis as a diagnostic test is not a reliable tool in the diagnosis of urinary tract infections (UTIs) in older adults or catheterized individuals (see pages 45–46 for a fuller discussion of this topic). With an ageing population, appropriate management of UTIs is an important focus of good care and AMS. Urine should not simply be tested due to its colour (i.e. appearing darker) or odour, as this may be more likely to indicate dehydration. Bacterial growth within the urinary tract is usually prevented by urinary flow, therefore poor fluid intake and dehydration can lead to increased risk of UTIs (Health Protection Scotland (HPS), 2018).

Key points to remember when obtaining samples are:

- Where possible, obtain samples before starting antibiotics (if not possible – DO NOT delay starting therapy).
- Comply with organizational policies and procedures for specimen collection.
- Obtain informed consent from the patient.
- Be competent in skill and knowledge about specimen collection, including understanding correct procedure and specimen handling.
- Use standard infection control precautions whilst obtaining samples.
- Include details of recent antimicrobial therapy on the sample laboratory request form.

Point-of-care testing

Rapid diagnostics and point-of-care testing (PoC) enable a wide range of tests to be carried out at a patient's bedside or within the community setting without the need for sophisticated laboratory technology (Price, 2001). In many clinical settings it is the nurse that carries out these tests and interprets the results to formulate a management plan for the patient. Rapid diagnostics and PoC are key AMS interventions as they enable rapid diagnosis of infection which, in turn, allows for appropriate antibiotic prescribing earlier in the patient journey. Both interventions have been included within national and global strategic plans as critical in preserving and effectively using antimicrobial therapies (World Health Organization (WHO), 2015; HM Government, 2019).

Interpreting results

Once samples have been sent to the laboratory they undergo sensitivity testing. With advances in technology, it is becoming increasingly common for this process to be automated, although traditional methods using agar plates are still used frequently. The purpose of culture and sensitivity is to grow and isolate causative organisms and identify the antimicrobial agents to which they are susceptible. Once the sample has been processed in the laboratory, an antibiogram is issued to the requesting clinician, even when results are negative. Antibiotics that will be effective in the treatment of infection are indicated with 'S' (sensitive) and those that are not are marked with 'R' (resistant). An example of a laboratory report is detailed in Table 4.1. Once the report is received, the nurse can communicate the results to the prescribing practitioner who, as part of a multi-professional discussion, can use the information to select the most appropriate antimicrobial treatment and route of administration, using the principles of antimicrobial stewardship, to determine the agent that is the narrowest spectrum and least likely to cause harm, i.e. resistance, Clostrioides difficile infection (CDI), toxicity.

Table 4.1. Example of laboratory report.

Sample: Blood Culture	Sensitivities	
Clinical details	Co-amoxiclav	R
Generally unwell and vomiting ?Sepsis	Gentamicin	S
Tests	Co-trimoxazole	S
Anaerobic	Amoxicillin	R
Aerobic	Aztreonam	S
Aerobic	Temocillin	R
Cultures	Pivmecillinam	S
Escherichia coli growth		
in both bottles		

Self-limiting infections

Differences in viral and bacterial infection presentation

Viral infections

- Most respiratory infections are caused by viruses.
- They are easily spread from person to person.
- The patient feels achy all over, with more than one site in the body affected.

Antibiotics will NOT cure viral infections. It is important to:

- treat with symptom relief, e.g. hydration and pain relief (community pharmacists can provide advice and guidance); and
- stop the spread – catch it, bin it, kill it!

Bacterial infections
These:

- are less common than viral infections;
- are not as easily spread from person to person;
- normally only affect one area of the body.

Antibiotics do work against bacterial infections. However, some bacterial infections are self-limiting in nature and do not require antibiotic treatment.

Respiratory tract infection
Most antibiotics are prescribed in primary care for respiratory tract infections (Gulliford *et al.*, 2014). In the UK, it has been estimated that 60% of antimicrobial prescriptions in primary care are for infections of the respiratory tract (National Institute for Health and Care Excellence (NICE), 2008). However, many of these infections are caused by viruses (NICE, 2008; Dekker *et al.*, 2019). Even when infections are caused by bacterial organisms, most will be self-limiting in nature and

will therefore not require antibiotics but will resolve with symptomatic relief and rest. The exception to this approach would be patients at risk of developing complications, including patients who are:

- systemically unwell;
- exhibiting signs and symptoms suggestive of serious illness and/or complications;
- at high risk due to pre-existing comorbidities;
- over 65 with acute cough and match two or more of the following criteria: hospitalized in the previous year; Type 1 or 2 diabetes; history of congestive heart failure; use oral glucocorticoids;
- over 80 with acute cough and match one or more of the above criteria.

In these populations, an immediate or back-up antibiotic prescription and/or further investigation should be offered (NICE, 2008).

Sore throat

The vast majority of throat infections are viral, with approximately 90% resolving within 7 days, without the need for antibiotics (Scottish Intercollegiate Guidelines Network (SIGN), 2010; British Medical Journal (BMJ), 2018; NICE, 2018a). Even those that are caused by bacterial infection will derive little benefit, in terms of symptom duration, from the use of antibiotics. Where antibiotics are used, illness duration is typically reduced by 12–18 h. Streptococci is the most likely causative organism in bacterial throat infection and only individuals with a high likelihood based on severity assessment require an antibiotic prescription.

The Centor score (Centor *et al.*, 1981) used to be the primary tool used to assess severity and likelihood of streptococcal infection. However, the FeverPAIN score (Little *et al.*, 2013) (see Table 4.2), more recently developed, provides greater accuracy with regards to which patients will benefit from an antibiotic.

Table 4.2. FeverPAIN score.

Fever in the last 24 hours		1 point
Purulence		1 point
Attend rapidly (under 3 days)		1 point
Inflamed tonsils		1 point
No cough or other respiratory symptoms		1 point
Total score		
Score	Likelihood of Streptococcal infection	Antibiotic prescribing
0–1	13–18%	No antibiotics required
2–3	34–40%	Consider 3-day back-up prescription
>4	62–65%	Immediate antibiotic prescription or short (<48 h) back-up prescription

All patients should be provided with advice about symptom relief and safety netting, irrespective of whether an antibiotic is prescribed.

Common cold

Viral infections such as the common cold or sinusitis can be as debilitating as bacterial infections. Healthcare workers can support patients with viral infections by providing information about the likely duration of symptoms, which for the common
cold can be up to 14 days (Mandell *et al.*, 2009; Thompson *et al.*, 2013), providing
reassurance about the lack of efficacy of antibiotics and providing advice on useful
self-care and symptom relief measures. Infection prevention and control advice
should also be offered to reduce the spread of infection. This advice can include:

*   covering the mouth when coughing/sneezing;
*   ideally coughing/sneezing into tissues;
*   discarding used tissues immediately;
*   increased hand hygiene (alcohol hand rub or hand washing), especially after
    coughing/sneezing or touching used tissues.

The Public Health England (PHE)/Royal College of General Practitioners
(RCGP) 'Treating Your Infection' leaflet (PHE, 2012) (Fig. 4.1) is a useful tool to use
when conducting a nursing assessment; it offers patients advice around symptom
control and safety netting and provides an opportunity to issue a back-up prescription. Back-up prescriptions can be provided as part of the safety netting process.
They are not intended to be dispensed immediately, but enable the patient to obtain antibiotics as per the direction of the issuing healthcare worker, should their
condition deteriorate or the severity persist for longer than expected.

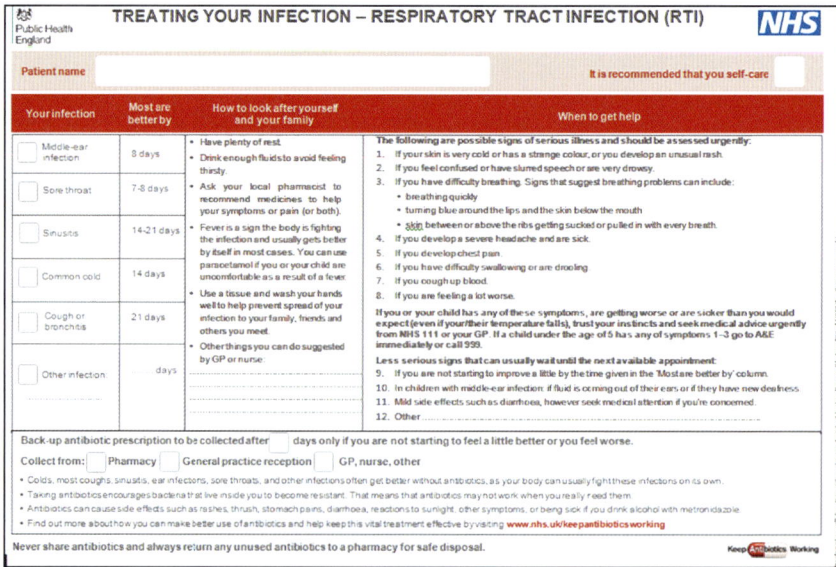

Fig. 4.1. Treating Your Infection leaflet.

Table 4.3. Differences in prescribing strategies in the treatment of respiratory infection. (From Spurling *et al.*, 2017)

	Patient satisfaction	Percentage of people who consumed antibiotics
Immediate prescription	92%	93%
Back-up prescription	87%	32%
No prescription	83%	14%

A review by Spurling *et al.* (2017) found that patient satisfaction with their consultation did not differ across prescribing strategies offered: immediate antibiotics versus back-up antibiotics versus no antibiotics in the treatment of respiratory infection. However, the greatest difference was seen in the consumption of antibiotics (Table 4.3), demonstrating that back-up strategies were an effective stewardship intervention. Evidence indicates that patient satisfaction with their consultation is increased when a comprehensive clinical assessment is carried out, information is provided on the likely duration of illness, explanation is provided around why antimicrobials are not required, reassurance is provided around illness, advice is given on symptom relief and safety netting instruction is provided should the illness deteriorate (Little *et al.*, 2013). The nursing role is integral to these steps.

Asymptomatic bacteriuria

As we age, our ability to maintain certain sterile environments within our body systems diminishes. This is particularly true of the bladder and urinary tract. What is normally a sterile environment in a healthy adult with no invasive devices *in situ* can become colonized with bacteria due to ageing, and when indwelling devices such as catheters are *in situ*. Colonization occurs when bacteria inhabit an area of the body and live there without causing harm (i.e. infection) to the individual (host). Individuals with long-term indwelling catheters will all have asymptomatic bacteriuria, often with two or more organisms. Table 4.4 describes the prevalence of asymptomatic bacteriuria.

Asymptomatic bacteriuria in older or catheterized individuals is often unnecessarily treated with antibiotics (SIGN, 2012; Zalmanovici Trestioreanu *et al.*, 2015; Rousham *et al.*, 2019), which increases the risk of AMR (Zalmanovici Trestioreanu *et al.*, 2015; Nicolle *et al.*, 2019; Rousham *et al.*, 2019), making future UTIs difficult to treat and increasing the risk of *Clostrioides difficile* infection. Evidence suggests that people who receive an antibiotic to treat a UTI are 2.5 times more likely to develop a drug-resistant organism than those who have not had antibiotic exposure (Costelloe *et al.*, 2010). According to a recent Cochrane review, patient outcomes (including the propensity to develop a UTI, complications and death) are not compromised if asymptomatic bacteriuria is not treated with antibiotics (Zalmanovici Trestioreanu *et al.*, 2015).

Table 4.4. Prevalence of asymptomatic bacteriuria. (From Colgan *et al.*, 2006; Zalmanovici Trestioreanu *et al.*, 2015)

Age	Women (%)	Men (%)
Premenopausal <50 years	1–5	n/a
Postmenopausal 50–70 years	3–9	n/a
Older community dwelling	15	4–19
Older long-term care-home residents	25–50	15–40

To Dip or Not to Dip – That Is the Question?

Urinalysis can be a good discriminatory screening tool for urinary infection in young adults (<65), pregnant women and children; however, it is not recommended to diagnose UTI in older adults or individuals with indwelling urinary catheters due to the increased prevalence of asymptomatic bacteriuria (SIGN, 2012; NICE, 2018b). A diagnosis of a UTI should be made following a person-centred assessment focusing on presenting signs and symptoms.

The Scottish Antimicrobial Prescribing Group (SAPG) has developed the algorithm below to assist with the diagnosis of UTI in older people (Fig. 4.2), including those with indwelling catheters. This is one example of a clinical resource developed to assist with minimizing the use of antibiotics and optimizing patient outcomes.

The Importance of Following Policy

An antimicrobial policy provides strategic guidance on the treatment of infection across all healthcare delivery settings, aiming to optimize patient outcomes whilst minimizing associated harm such as resistance, toxicity and clostrioides difficile infection (CDI) (British Society for Antimicrobial Chemotherapy (BSAC), 2018). Antimicrobial guidelines are evidence-based and incorporate local resistance patterns to inform prescribing decisions. When compiling an antimicrobial policy, spectrum of activity, efficacy in treating infection, toxicity to patient, likelihood of promoting resistance, accessibility and cost are taken into consideration. Adherence to antimicrobial policy ensures the application of evidence-based medicine and offers patients the safest and most effective treatment for their infection. Antimicrobial policies are viewed as a critical stewardship intervention as they are intended to ensure consistency of antimicrobial use across clinical practice settings.

Antimicrobial Policies and Nursing Practice

There is a perception that antimicrobial policies/guidelines are solely for the purpose of prescribers; however, antimicrobial policies can facilitate and support the

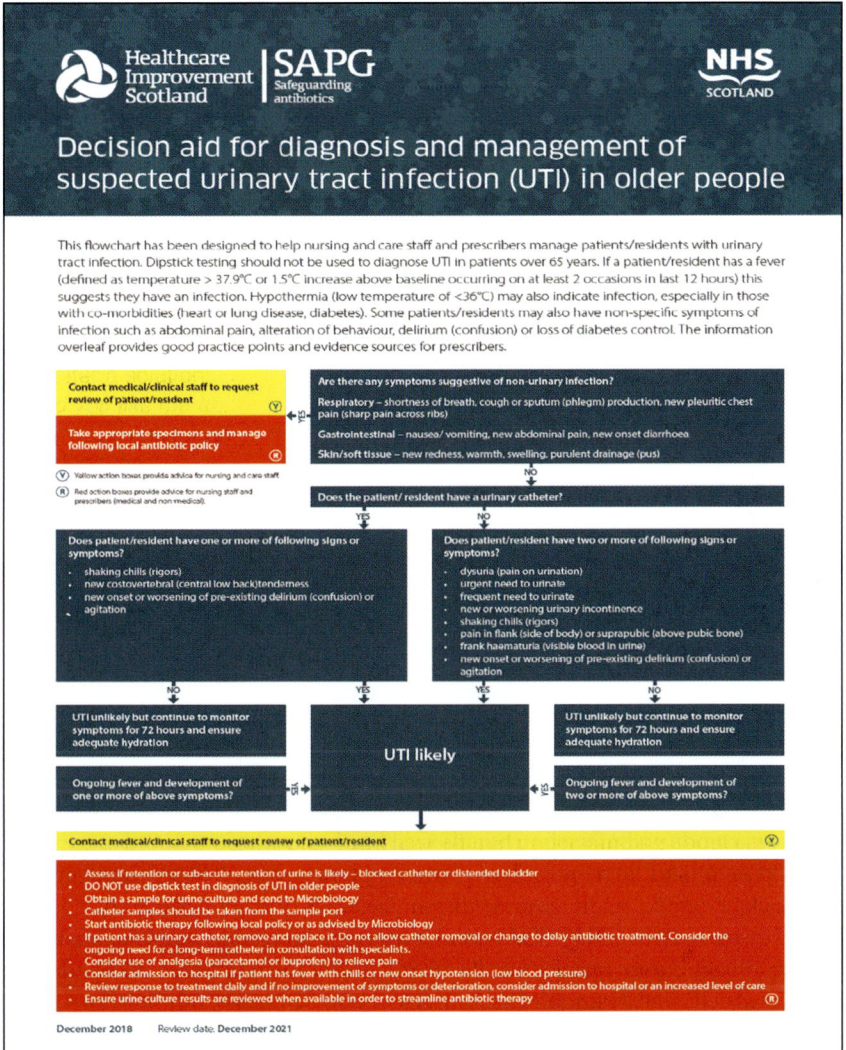

Fig. 4.2. Decision aid for the diagnosis and management of suspected urinary tract infection in older people. (From SAPG, 2018)

role of the administrant. When considering the information that is required prior to administering any medicine, such as indication, drug choice, dose, dose frequency and duration, all of this is featured within the antimicrobial guidelines. Nurses can use antimicrobial guidelines to fulfil their registrant duties in the safe and effective administration of medicines, thus optimizing patient care outcomes. Furthermore, the use of antimicrobial guidelines in combination with clinical assessment of patients can determine response to therapy and identify opportunities for multi-disciplinary conversations. This may include IV-to-oral switch, de-escalation of therapy, targeted therapy as per laboratory sensitivities, stopping antibiotic therapy, promoting appropriate duration or escalating therapy.

The role of antimicrobial stewardship is not exclusive to the registered nurse; it is also relevant to care assistants, who are often responsible for personal care delivery, assessment of patients and sample collection. Additionally, antimicrobial stewardship can be embedded within newly evolved roles, such as Nursing Associates, in England, who have a medicines administration role.

Nurses in the community and out-of-hours practice are often involved in administering or supplying antibiotics under the instruction of a Patient Group Direction (PGD). PGDs are written instructions developed by a multi-professional team in the treatment of specific diseases/indications. Examples of infections that are frequently managed in these care settings using PGD are UTI, cellulitis, bites (human and animal) and sore throat.

Nurse Prescribers and Policy

In the UK there is a growing number of nurse prescribers, with figures set to continue to rise as newly qualified practitioners will graduate as prescriber-ready from 2023. Undergraduate education will change to incorporate the pharmacological element of the prescribing qualification, which will enable newly qualified practitioners to undertake the practical element once qualified and aligned to clinical service need. Evidence suggests that nurse prescribers understand the expectations of patients who have an infection and use a patient-centred approach when they decide to apply management strategies that do not include antibiotics (Courtenay *et al.*, 2016). Nurse prescriber adherence to guidelines is well documented with the ten most commonly prescribed antibiotics being those referred to in guidelines and accounting for 95.8% of all nurse prescriptions in primary care settings (HPS, 2018).

The landscape of care delivery is changing, and so too is the likelihood that nurses, whether they are prescribers or not, will continue to influence optimal antimicrobial prescribing.

AWaRe Antibiotics and Restriction Principles

In 2017, the World Health Organization published their most significant revision in 40 years of the antibiotic section of the Essential Medicines List. Rebranded as AWaRe, antibiotics were classified into three categories: Access, Watch and Reserve, with recommendations on when antibiotics should be used within each category. This stratification of antibiotics aims to promote optimal treatment outcomes for patients, minimize the development of AMR, preserve effectiveness for future generations and support the strategic goals of the 2015 WHO Global Action Plan (GAP). The five strategic objectives of the WHO GAP are:

- Improve awareness and understanding of AMR;
- Enhance knowledge through research and surveillance;
- Minimize the incidence of infection;
- Optimize the use of antimicrobial agents;
- Develop an economic case for sustainable investment in new medicines, diagnostic tools, vaccines and alternative interventions, accounting for the needs of all countries.

In the UK, Public Health England has further refined this list of AWaRe antibiotics (Table 4.5) that should be promoted across all healthcare delivery settings.

Antibiotic restriction (limiting and/or controlling the antibiotics a clinician can prescribe) is identified as a key stewardship strategy and has been associated with reduced antibiotic consumption and AMR (Hernandez-Santiago *et al.*, 2019).

Table 4.5. AWaRe list.

Access	Watch	Reserve
Amoxicillin/ampicillin	Amicacin/tobramycin etc.	Aztreonam
Penicillin (all forms)	Macrolides	Ceftobiprole
Co-trimoxazole	Most cephalosporins	Ceftaroline
Doxycycline	Chloramphenicol	Ceftazidime/avibactam
Flucloxacillin	Fluoroquinolones	Ceftolozane/tazobactam
Fosfomycin oral	Clindamycin	Colistin
Fusidate	Co-amoxiclav	Daptomycin
Gentamicin	Other tetracyclines	Carbapenems
Metronidazole	Fidaxomycin	Fosfomycin IV
Nitrofurantoin	Piperacillin/tazobactam	Linezolid/tedizolid
Pivmecillinam	Temocillin	Televancin
Tetracycline	Vancomycin, teicoplanin	Tigecycline
Trimethoprim		

Choosing the Antibiotic and Documentation

The prevalence of AMR has risen over the last 40 years and very few novel antimicrobials have been developed. This has led to increased pressure on existing antibiotics and greater challenges in treating patients (PHE, 2015). Inappropriate use of antimicrobials increases the risk to patients of colonization and infection with resistant organisms, and subsequent transmission to other patients (PHE, 2015). Since bacteria evolve to resist the effects of existing antibiotics, infections can become more difficult to treat and, subsequently, modern medical interventions can become more dangerous for patients (Wells and Piddock, 2017).

Due to this increased pressure on antimicrobial sustainability, healthcare workers should be aware of ways to reduce unnecessary and inappropriate prescribing. Antimicrobial stewardship is defined as 'the right antibiotic, for the right indication, the right patient, at the right time with the right dose and route' (BSAC, 2018). It is by following these principles that stewardship will help to address the global issue of AMR.

The growth of micro-organisms can be inhibited by many different chemical compounds. The term 'antibacterial' or 'antibiotic' is used to describe an agent that is active against bacteria only and has been licensed for the treatment of infection in humans or animals. Antibiotics work by interfering with the bacterial cell wall, which can prevent the bacteria from growing and replicating, or kill the bacteria altogether (see Chapter 3). There are several different classes of antibiotics, all of which interfere with these critical mechanisms of cellular replication in different ways. Antibiotics have the potential to kill Gram-positive or Gram-negative bacteria or both. This latter group (i.e. those that kill both) are called broad-spectrum antibiotics. Increased use and especially inappropriate use of broad- and very broad-spectrum agents such as carbapenems and piperacillin/tazobactam can select organisms with resistance (Robson *et al.*, 2018).

When deciding whether or not to prescribe an antibiotic, the prescriber needs to balance the benefit to the patient with the potential harm that may be caused by side effects, microbiota changes and the societal harm caused by exposure to antibiotics and the risk of antimicrobial resistance. There are several parameters that should be considered before prescribing any antibiotic to ensure prudent prescribing whilst optimizing patient outcome. The route by which an antibiotic is delivered is extremely important both for the patient and for addressing the stewardship focus. There are some serious infections, e.g. endocarditis and bone and joint infections, that require intravenous (IV) therapy due to the penetration and absorption capability of oral agents. A process for reviewing all patients on IV antibiotics and assessing each patient's eligibility for conversion to the oral route should be mandated in all patients. The benefits of oral therapy over IV therapy are well documented in the literature and include reduced exposure to resistant pathogens through the IV site, reduced incidence of soft skin and tissue infection and blood stream infection, reduced preparation and administration time for nursing staff and, potentially, a decreased length of stay for the patient within the healthcare facility.

Correct duration of therapy is essential to ensure that the infection is treated adequately but also that adverse effects are minimized. Most common infections will be treated with antibiotics for a maximum of 7 days. Extending antibiotic therapy beyond this increases the risk of emerging resistant strains (Wald-Dickler and Spellberg, 2019). Evidence is now beginning to emerge suggesting that 5 days' duration will replace the standard 7-day course (Wald-Dickler and Spellberg, 2019). Good documentation of infection diagnosis and management should be promoted at the time of antibiotic initiation. Intravenous antimicrobial prescriptions should be reviewed at 48–72 hours in all health and care settings (including community and outpatient services), including response to treatment and microbiological results, to determine if the antimicrobial needs to be continued and, if so, whether it can be switched to an oral antimicrobial (NICE, 2015). The duration of therapy should be recorded on the patient's drug administration chart and the medical notes. This process of documenting the patient's management plan is imperative to ensure communication across all healthcare professionals and to promote a seamless level of care.

Timing of delivery is more important for antibiotics than for many other medicines because antibiotics are usually administered several times per day. Timely administration and avoidance of missed doses are two key contributions to antimicrobial stewardship. The time between doses affects whether or not the invading organism is killed and also influences the development of resistant strains. Prompt administration of antibiotics is essential in patients diagnosed with sepsis, as each hour of delay significantly reduces survival (Mi *et al.*, 2019).

Antibiotic allergy is another factor that has to be considered when choosing an appropriate therapy. Antibiotics are the commonest cause of life-threatening drug reactions, including anaphylaxis and organ-specific reactions; however, many antibiotic reactions documented as allergies are unknown or not remembered by the patient (NICE, 2014). Although such reactions pose negligible risk to patients, they currently represent a global threat to public health. Antibiotic allergy labels result in displacement of first-line therapies for antibiotic prophylaxis and treatment (Blumenthal *et al.*, 2019). A penicillin allergy label, in particular, is associated with increased use of broad-spectrum antibiotics, which results in increased adverse events and antibiotic resistance. Most patients labelled as allergic to penicillins are not allergic when appropriately stratified for risk, tested and re-challenged (Blumenthal *et al.*, 2019).

As with any medicine, antimicrobials can have side effects. The most common side effect includes upset of the gastrointestinal tract such as nausea, vomiting and diarrhoea. Table 4.6 describes the side effects of commonly used antibiotics.

Table 4.6. Examples of side effects of commonly used antibiotics[*]. (From British National Formulary, 2019)

Drug	Common side effect
Amoxicillin	Hypersensitivity/skin reactions
Ciprofloxacin	*C. difficile* infection, tendon rupture, seizures
Gentamicin	Nephro/oto-toxicity
Doxycycline	Photosensitivity, stains teeth
Metronidazole	Metallic taste in mouth, nausea
Trimethoprim	Blood dyscrasias
Vancomycin	'Red man' syndrome (when infused too quickly)

[*]This list is by no means exhaustive – please refer to local/national prescribing formulary for full details.

Antimicrobial Stewardship in Action

Case study 1

Christine is a 76-year-old woman who attended the primary care centre as she experienced episodes of dizzy spells. The practice nurse carried out a clinical assessment. Christine's observations were: blood pressure 102/51, pulse 73, respiratory rate 16, temperature 37.2°C. The nurse asked Christine to provide a midstream urine sample. A urinalysis was carried out which showed:

- Leucocytes, nitrite & protein – positive
- Glucose, blood & ketones – negative

Christine did not have any significant previous clinical history other than mild-urinary incontinence, which she has had for 20 years, and the only medication she was taking was paracetamol for hip pain.

The clinical assessment was communicated to the general practitioner who started a course of antibiotics for a suspected UTI.

Question 1: Does Christine have a urinary tract infection?
Answer: Christine does not have any signs or symptoms that are indicative of a urinary tract infection.

Question 2: Is there any benefit to carrying out urinalysis in patients over the age of 65?
Answer: Many older adults have bacteria living in their bladder causing no harm – this is known as 'asymptomatic bacteriuria'.

When bacteria break down, they produce nitrites. The presence of bacteria promotes an increase in white cells (leucocytes); therefore, urinalysis is not indicated in the diagnosis of UTI in older adults. Diagnosis should be made on presenting signs and symptoms, making this a more patient-focused assessment.

Question 3: What are the risks associated with unnecessary antibiotic use?
Answer: Christine was prescribed a course of antibiotics that she did not need. This is patient harm. Patients who are treated with a single course of antibiotics for UTI are 2.5 times more likely to develop a resistant organism. This has the potential to complicate treatment if they do develop a UTI in the future.

Question 4: What alternative lifestyle measures could be promoted with Christine to minimize her risk of actually developing a UTI?
Answer:
- adequate hydration
- avoid constipation
- good toilet hygiene (front to back for ladies)
- adequate bladder voiding
- continence management
- catheters only if absolutely necessary
- standard infection control precautions (SICPs)

Case study 2

David is a 69-year-old man admitted to hospital with cellulitis of the right leg. On admission, his observations suggested he was septic and he was commenced on intravenous flucloxacillin. By day 2 of the therapy, his observations returned to normal parameters and his clinical picture was improving. You come on shift on day 16 of David's admission to discover that he is still on IV flucloxacillin.

Question 1: Is this an appropriate duration of therapy for cellulitis?
Answer: Most uncomplicated soft skin and tissue infections only require 7 days IV/PO antibiotic therapy. Therefore, in David's case, he has received 9 days of unnecessary antibiotic therapy.

Question 2: Did David require IV antibiotics for the entire duration of therapy?
Answer: David did not require IV antibiotics for the entire duration of his therapy. As soon as a person is apyrexial, showing clinical signs of improvement (vital signs returning to normal and CRP, WBC improving – CRP and WBC don't necessarily need to be within normal parameters just on a downward trajectory), the oral route is available. So long as the person does not have an infection that specifically requires IV therapy (i.e. meningitis, osteomyelitis, abscess, endocarditis) IV therapy can be changed to the oral route.

IV antibiotics must be reviewed on a daily basis – points to consider during review: clinical need for IV: Are sensitivities available from microbiology? Can the number of antibiotics be reduced? Nursing staff can contribute to the decision to de-escalate therapy as they have a frontline role in the clinical assessment of patients and monitor oral intake. Follow IVOST guidelines for all patients receiving IV antibiotics.

Question 3: At what point in care could an intervention have been made?
Answer: David could have been changed to oral therapy on day 2 when his observations returned to normal parameters.

Question 4: What are the associated risk factors for David continuing antibiotics for this duration?
Answer:

14 days of unnecessary PVC insertions resulting in:

- potential to develop skin and soft tissue infection at PVC entry site;
- potential to develop bloodstream infection through extended PVC use.

9 days of unnecessary antibiotic therapy resulting in:

- potential to develop resistant organisms;
- potential to develop CDI.

Key Points

1. Nurses are an important and highly valued component of the antimicrobial stewardship team who are integral to all aspects of patient care.
2. Local guidance on infection prevention and control and antimicrobial treatment ensures the application of evidence-based medicine whilst ensuring patients receive the safest and most effective treatment for their infection.
3. All healthcare workers should follow the principles of antimicrobial stewardship to help address the global issue of antimicrobial resistance.
4. Appropriate sampling is key to diagnosing infection and tailoring antimicrobial therapy to the needs of the patient based on their clinical signs and symptoms.
5. Antibiotics are not required in the management of self-limiting bacterial infections and are ineffective in the management of viral infections (e.g. the common cold and some respiratory infections).

Further Reading

ANA (American Nurses Association) (2017) Redefining the antibiotic stewardship team: recommendations from the American Nurses Association/Centers for Disease Control and Prevention Workgroup on the role of registered nurses in hospital antibiotic stewardship practices. Silver Springs, American Nurses Association.

Dougherty, L. and Lister, S. (2015) *Royal Marsden Manual of Clinical Nursing Procedures*, 9th Edition. Royal Marsden NHS Foundation Trust.

FutureLearn Antimicrobial and Antibiotic Resistance Courses. Available at: https://www.futurelearn.com/courses/categories/health-and-psychology-courses/antimicrobial-and-antibiotic-resistance (accessed 12 December 2019).

NHS Education for Scotland (2019) *Raising Awareness of Antimicrobial Stewardship for Nurses and Midwives.* NES, Edinburgh. Available at: https://www.nes.scot.nhs.uk/education-and-training/by-theme-initiative/healthcare-associated-infections/training-resources/raising-awareness-of-antimicrobial-stewardship-for-nurses-and-midwives.aspx (accessed 20 December 2019).

Public Health England (2012) *TARGET Antibiotics Toolkit*. Royal College of General Practitioners. Available at: http://rcgp.org.uk/targetantibiotics (accessed 16 April 2019).

References

Bentley, J., Thakore, S., Muir, L., Baird, A. and Lee, J. (2016) A change of culture: reducing blood culture contamination rates in an emergency department. *BMJ Quality Improvement Report* 5(1).

Blumenthal, K., Peter, J., Trubiano, J. and Philips, E. (2019) Antibiotic allergy. *The Lancet* 393, 183–198.

BMJ (2018) *BMJ Best Practice: Acute Pharyngitis*. BMJ, London.

BNF (British National Formulary) (2019) BMJ Group. Pharmaceutical Press, London.

BSAC (British Society for Antimicrobial Chemotherapy) (2018) *Antimicrobial Stewardship: From Principles to Practice*, 1st Edition. BSAC, Birmingham, UK.

Centor, R.M., Witherspoon, J.M., Dalton, H.P., Brody, C.E. and Link, K. (1981) The diagnosis of strep throat in adults in the emergency room. *Medical Decision Making* 1(3), 239–246. DOI: 10.1177/0272989X8100100304

Colgan, R., Nicolle, L., McGlone, A. and Hooton, T.M. (2006) Asymptomatic bacteriuria in adults. *American Family Physician* 74(6), 985–990.

Costelloe, C., Metcalfe, C., Lovering, A., Mant, D. and Hay, A.D. (2010) Effect of antibiotic prescribing in primary care on antimicrobial resistance in individual patients: systematic review and meta-analysis. *British Medical Journal* 340, c2096.

Courtenay, M., Owbotham, S., Lim, R., Deslandes, R., Hodson, K., *et al.* (2016) Antibiotics for acute respiratory tract infections: a mixed-methods study of patient experiences of non-medical prescriber management. *BMJ Open* 7.

Dekker, A.R.J., van der Velden, A.W., Luijken, J., Verheij, T.J.M. and van Giessen, A. (2019) Cost-effectiveness analysis of a GP- and parent-directed intervention to reduce antibiotic prescribing for children with respiratory tract infections in primary care. *Journal of Antimicrobial Chemotherapy* 74, 1137–1142.

Gulliford, M.C., van Staa, T., Dregan, A., McDermott, L., McCann, G., *et al.* (2014) Electronic health records for intervention research: a cluster randomized trial to reduce antibiotic prescribing in primary care. *Annals of Family Medicine* 12(4), 344–351. DOI: 10.1370/afm.1659

Health Protection Scotland (2018) *Scottish One Health Antimicrobial Use and Antimicrobial Resistance Report 2017*. Health Protection Scotland, Glasgow, UK.

Hernandez-Santiago, V., Davey, P.G., Nathwani, D., Marwick, C.A. and Guthrie, B. (2019) Changes in resistance among coliform bacteraemia associated with a primary care antimicrobial stewardship intervention: a population-based interrupted time series study. *PLOS Medicine* 16(6). Available at: https://doi.org/10.1371/journal.pmed.1002825 (accessed 19 September 2019).

HM Government (2019) *Tackling Antimicrobial Resistance 2019–2024. The UK's Five-year National Action Plan*. HM Government, London.

Inglis, T.J.J. (2003) *Microbiology and Infection*, 2nd Edition. Churchill Livingstone, Philadelphia, Pennsylvania.

Little, P., Moore, M., Hobbs, F.D.R., Mant, D., McNulty, C. *et al.* (2013) PRImary care Streptococcal Management (PRISM) study: identifying clinical variables associated with Lancefield group A β-haemolytic streptococci and Lancefield non-group A streptococcal throat infections from two cohorts of patients presenting with an acute sore throat. *BMJ Open* 3(10).

Mandell, G., Bennett, J. and Dolin, R. (eds) (2009) *Mandell, Douglas and Bennett's Principles and Practice of Infectious Diseases*, 7th Edition, Churchill Livingstone, Philadelphia, Pennsylvania.

Mi, M.Y., Klompas, M. and Evans, L. (2019) Early administration of antibiotics for suspected sepsis. *New England Journal of Medicine* 380(6), 593–596.

NHS Education for Scotland (2017) *Taking a Blood Culture Sample* [homepage of NHS Education for Scotland]. Available at: https://www.nes.scot.nhs.uk/education-and-training/by-theme-initiative/healthcare-associated-infections/online-short-courses/aseptic-technique.aspx.

NICE (National Institute for Health and Care Excellence) (2008) *Respiratory Tract Infections (Self-limiting): Prescribing Antibiotics* [CG69]. NICE, London.

NICE (2014) *Drug Allergy: Diagnosis and Management* [CG183]. NICE, London.

NICE (2015) *Antimicrobial stewardship: systems and processes for effective antimicrobial medicine use*. [NG15]. NICE, London.

NICE (2018a) *Sore Throat (Acute): Antimicrobial Prescribing*. NICE guideline [NG84]. NICE, London.

NICE (2018b) *Urinary Tract Infection (Lower): Antimicrobial Prescribing* [NG109]. NICE, London.

Nicolle, L.E., Gupta, K., Bradley, S.F., Colgan, R., DeMuri, G.P., *et al*. (2019) Clinical practice guideline for the management of asymptomatic bacteriuria: 2019 update by the Infectious Diseases Society of America. *Clinical Infectious Diseases* 68, e83–e110.

PHE (2012) *TARGET Antibiotics Toolkit*. Royal College of General Practitioners. Available at: http://www.rcgp.org.uk/targetantibiotics (accessed 16 April 2019).

PHE (2015) *Start Smart – Then Focus: Antimicrobial Stewardship Toolkit for English Hospitals*. Public Health England, London.

Price, C.P. (2001) *Point of Care Testing*. British Medical Journal Publishing Group, London.

Raja, N., Parratt, D. and Meyers, M. (2009) Blood culture contamination in a district general hospital in the UK: a 1-year study. *Healthcare Infection* 14, 95–100.

Robson, S.E., Cockburn, A., Sneddon, J., Mohana, A., Bennie, M., *et al*. (2018) Optimizing carbapenem use through a national quality improvement programme. *Journal of Antimicrobial Chemotherapy* 73(8), 2223–2230.

Rousham, E., Cooper, M., Petherick, E., Saukko, P. and Oppenheim, B. (2019) Overprescribing antibiotics for asymptomatic bacteriuria in older adults: a case series review of admissions in two UK hospitals. *Antimicrobial Resistance and Infection Control* 8(1).

SAPG (Scottish Antimicrobial Prescribing Group) (2018) *Decision Aid for Diagnosis and Management of Suspected Urinary Tract Infection (UTI) in Older People* [homepage of Health Improvement Scotland]. Available at: https://www.sapg. scot/media/4092/uti-in-older-people.pdf (accessed 19 September 2019).

SIGN (2010) *SIGN 117. Management of Sore Throat and Indications for Tonsillectomy: A National Clinical Guideline*. Health Improvement Scotland, Glasgow, UK.

SIGN (2012) *SIGN 88. Management of Suspected Bacterial Urinary Tract Infection in Adults: A National Clinical Guideline*. Health Improvement Scotland, Glasgow, UK.

Spurling, G.K.P., Del Mar, C.B., Dooley, L., Foxlee, R. and Farley, R. (2017) *Delayed antibiotic prescriptions for respiratory infections (review). Cochrane Database of Systematic Reviews*.

Thompson, M., Vodicka, T.A., Blair, P.S., Buckley, D.I., Heneghan, C. and Hay, A.D. (2013) *Duration of Symptoms of Respiratory Tract Infections in Children: Systematic Review*. British Medical Journal Publishing Group, London.

Wald-Dickler, N. and Spellberg, B. (2019) Short-course antibiotic therapy – replacing Constantine Units with 'Shorter Is Better'. *Clinical Infectious Diseases*. DOI: 10.1093/cid/ciy1134

Wells, V. and Piddock, L.J.V. (2017) Addressing antimicrobial resistance in the UK and Europe. *The Lancet Infectious Diseases* 17, 1230–1231.

WHO (World Health Organization) (2015) *Global Action Plan on Antimicrobial Resistance*. WHO, Geneva, Switzerland.

WHO (2017) *WHO Model List of Essential Medicines*, 20th Edition. WHO, Geneva, Switzerland.

Zalmanovici Trestioreanu, A., Lador, A., Sauerbrun-Cutler, M.T. and Leibovici, L. (2015) *Antibiotics for asymptomatic bacteriuria (review). Cochrane Database of Systematic Reviews*.

Antimicrobial Prescribing Practice

5

Enrique Castro-Sánchez

Lead Academic Research Nurse, NIHR HPRU
In Healthcare Associated Infection and Antimicrobial
Resistance, Imperial College, London

Objective: To be aware of how antimicrobials are used in practice in terms of their dose, timing, duration and appropriate route of administration, and apply this knowledge to nursing practice.

Recognizing and Managing Sepsis

Sepsis is defined as a life-threatening organ dysfunction caused by a dysregulated host response to infection (Singer *et al.*, 2016). Worldwide, more than 30 million people are affected by sepsis every year, with up to 6 million deaths (Fleischmann *et al.*, 2016). In the UK, sepsis led to 77,996 admissions and 15,851 deaths in 2016/17 (NICE, 2016).

Septic shock is the most severe presentation of sepsis, with circulatory and cell/metabolic dysfunction associated with a higher risk of mortality, and often featuring a drop in systolic blood pressure that does not respond to fluid resuscitation (NICE, 2016). Patients with septic shock can be identified using the following criteria: persistent hypotension requiring vasopressors to maintain mean arterial pressure (MAP) ≥65 mm Hg, and serum lactate >2 mmol/l (or 18 mg/dl) despite adequate volume resuscitation (Singer *et al.*, 2016). Organ dysfunction is defined as an increase in the Sequential (Sepsis-related) Organ Failure Assessment (SOFA) score, with a score of 2 points or more. The SOFA score calculates the level of dysfunction in five systems, namely respiratory, cardiovascular, coagulation, renal and neurologic. This score uses bedside clinical variables and calculates morbidity rather than mortality (Vincent *et al.*, 1996). As well as the SOFA score, the quick Sequential Organ Failure Assessment (qSOFA), a bedside index which can be used outside critical care units, can be used to identify patients with suspected

Email: e.castro-sanchez@imperial.ac.uk

> **Box 1.** 'Hour-1' sepsis interventions.
> _____
>
> - measuring lactate levels and remeasuring if the initial level is >2 mmol/l;
> - obtaining blood cultures prior to the administration of antimicrobials;
> - administration of broad-spectrum antimicrobials;
> - rapid administration of 30 ml/kg crystalloid for hypotension or lactate ≥4 mmol/l; and
> - vasopressors if the patient is hypotensive during or after fluid resuscitation to maintain MAP above 65 mm Hg.

infection who are likely to develop sepsis. The qSOFA index includes alteration in mental status (Glasgow Coma Scale <15), systolic blood pressure ≤100 mm Hg, and respiratory rate ≥22/min.

For the diagnosis of sepsis, appropriate routine microbiologic cultures (including two sets of blood cultures, aerobic and anaerobic) should be collected as soon as possible before antimicrobial therapy is commenced (Rhodes *et al.*, 2016).

Resuscitation and management of patients should start immediately, with current guidelines recommending the 'Hour-1 bundle', a quality improvement intervention focused on recognizing the medical emergency that sepsis is, and commencing several interventions in the first hour following sepsis recognition (Levy *et al.*, 2018; Box 1).

Nurses can play a crucial role in the prompt identification of patients with sepsis, in view of their close and constant focus on care. As a result of such attention and close relation, several screening interventions have embraced nurses to effect improvements. For example, protocols defining roles and tasks for nurses related to sepsis, including prompt obtention of blood cultures and lactate, as well as administering fluid repletion, resulted in improvements in the time to fluid resuscitation, administration of antibiotics and collection of lactate levels (Coates *et al.*, 2015).

Similar bundled-based interventions have achieved comparable results, including the implementation of protocols that formalize measurement of serum lactate, obtention of blood cultures prior to antibiotic therapy and prompt administration of antibiotics to patients attending the emergency department (Tromp *et al.*, 2010; Kumar *et al.*, 2015). In addition to the components included in the Hour-1 bundle, these quality improvement interventions have also provided educational sessions, decision support charts, posters, audit and feedback.

The Use of Guidelines to Initiate Prompt and Effective Antimicrobial Treatment

Prescription and management of antimicrobials may be one of the most dynamic clinical areas for nurses. The sustained development of antimicrobial resistance, the emergence of new and re-emergence of previously common infections, as well as the development and availability of new drugs, demand continued professional development in the area of antimicrobial stewardship. Additionally, as infections can affect patients in any clinical setting, it is guaranteed that nurses in every speciality will need to know how to prepare, administer and manage antimicrobials, including how to participate in and influence the interdisciplinary decisions about antimicrobial use (Broom *et al.*, 2017).

Guidelines are among the components often included within antimicrobial stewardship quality improvement interventions (NICE, 2015), in common with many other areas of clinical practice, in an attempt to standardize antibiotic treatment decisions and improve clinical outcomes for patients whilst optimizing the use of resources. Ideally, guidelines will be based on the best available evidence (Chambers *et al.*, 2019), examined by experts and other relevant stakeholders, and allowing for patient preferences to be weighted in among the therapeutic options. Nurses should proactively engage with this process of antimicrobial guideline development, implementation and evaluation, as part of interdisciplinary work and as key clinical stakeholders in AMS interventions.

Whilst guidelines can lead to better decisions about antibiotic prescribing and management, they are not exempt from flaws (Fitzpatrick *et al.*, 2019). Essentially, these issues can be framed within broader challenges to the notion of evidence-driven practice, challenges to the fidelity of the implementation and use of guidelines in such clinical practice, and tensions between the therapeutic options preferred by the guidelines and the local clinical reality.

For example, some authors have already highlighted some shortcomings of evidence-based practice (Greenhalgh *et al.*, 2014), with concerns about influences on the evidence used and the process of deciding how to use it, the unmanageable amount of information available, the tensions between statistical and clinical benefits, and the lack of guidelines fitness to respond to multimorbidity. As advocates for patients, nurses should be aware of these limitations, ensuring that recommendations included within any documents are relevant to patients due to be treated, and enabling the input of patient preferences to the decision-making process (see Chapter 6).

Secondly, the implementation of guidelines and their recommendations, including their fidelity (i.e. how accurately the use in the real world mimics the proposed use) (Rzewuska *et al.*, 2019), as well as the applicability of the guidelines to a specific local context, may result in successful or failed adoption of such guidance. As mentioned, guidelines may not be relevant to the local context, which may be difficult to define. For example, the study by Moore *et al.* (2014) reported on resistant organisms isolated from patients in various clinical departments in a large hospital in London. The findings clearly demonstrated the high prevalence and characteristics of drug-resistant pathogens in the local ICU, compared with the rest of the wards. If the hospital-wide antimicrobial policy were to be followed up without consideration to local (in this case ward-level) patterns of resistance, many patients would receive an ineffective antibiotic. For such reason, local guidelines should be informed by local data, as granular as possible, including local sensitivity, pathogens and clinical outcomes of therapy. Using antimicrobial therapy empirically (i.e. whilst the results of biological samples such as blood cultures are returned) (Chiotos *et al.*, 2019) remains a challenge for clinicians, including nurses, as a result of the gap between antibiotics recommended by guidelines and local susceptibility patterns.

To sum up this section, it is paramount for nurses not only to know about any existing national, organizational or local antibiotic guidelines, but also to understand the potential shortcomings those documents may have, and proactively participate in the process of writing and reviewing the treatment policies.

Guidance on Completing a Course of Antimicrobials

An essential role for all nurses worldwide involves supporting patients to take medication appropriately, which, in the case of antibiotics, includes completing the prescribed course for the duration of doses or days indicated. However, the matter of optimal duration of antimicrobial courses has received increasing attention in recent years, as reducing unnecessary antimicrobial exposure would be beneficial to mitigate drug-resistant infections at both individual (Costelloe *et al.*, 2010; Fisher *et al.*, 2018) and population level (Pouwels *et al.*, 2018). Such reduction seems even more necessary in light of the data regarding excessive and prolonged duration of antimicrobial courses reported by Pouwels *et al.* (2019) in their evaluation of alignment of antibiotic prescriptions with recommended guidelines in primary care. The analysis by Llewelyn *et al.* (2017), additionally, argued that the evidence regarding optimal antibiotic duration, clinical effectiveness and resistance was lacking, and that harm to patients is more likely to originate from antimicrobial resistance derived from unnecessarily long courses of antibiotics than it is from stopping them early. This viewpoint also stressed that the evidence about stopping antibiotic treatment, leading to increased risk of resistant infection, was lacking. As healthcare workers administering medication to patients (and, in some cases, depending on local legislation, also prescribing antibiotics), nurses should be familiar and engage with this debate so that they can respond to any concerns from patients and families. Furthermore, there is already increasing evidence supporting recommendations for shorter antibiotic courses (Wilson *et al.*, 2019; Table 5.1).

For instance, the systematic review by Dawson-Hahn *et al.* in 2017 found similar cure rates and relapse when 3–7 days of antibiotics were given for acute bacterial sinusitis compared to 6–10 days. Equally, another systematic review (Kozyrskyj *et al.*, 2010) favoured a short antibiotic course for acute otitis media in children. For uncomplicated urinary tract infections, for example, there was no significant difference in cure rates when women received 3 days of antibiotics compared with courses of 5 days or longer (Milo *et al.*, 2005).

Table 5.1. Recommendations for antibiotic course duration. (Adapted from Wilson *et al.*, 2019)

Diagnosis	Indications for antibiotic therapy	Duration
Acute tonsillopharyngitis	2–25 years, high risk of acute rheumatic fever, or rheumatic heart disease, or scarlet fever	10 days
Acute rhinosinusitis	Symptoms >7 days, or high fever >3 days, or biphasic illness	5 days
Acute otitis media	<6 months old, or systemic symptoms	5 days 7 days
Community-acquired pneumonia (mild, can review progress in 48 h)	-	5–7 days 3–5 days 3–5 days 3–5 days
Uncomplicated urinary tract infection	-	3 days 5 days 7 days 3–5 days
Cellulitis (mild, low risk for MRSA)	-	5 days 5 days
Impetigo	-	7 days 3–10 days 5 days single dose
Abscess (low risk for MRSA)	Spreading cellulitis, or systemic symptoms, or large lesion/ critical area	5 days

Switching from Intravenous Antimicrobials to Oral Therapy

Optimal administration and management of intravenous antimicrobials remains a cornerstone of antibiotic-related practice and offers an opportunity for nurses to combine several key patient safety interventions focused on vascular access care, infection prevention and control, and AMS. The importance of this constituent of medicines management is, for example, recognized by national guidance about optimal AMS behaviours such as 'Start Smart Then Focus' (Public Health England (PHE), 2015) in the UK, stressing the need to proactively evaluate the clinical improvement of patients, establishing the need to continue intravenous antibiotics beyond 48–72 hours.

Several criteria need to be met before intravenous antimicrobials can be switched to the oral route. These criteria include, for example: a temperature

<38°C for a period of 24 h; signs and symptoms of infection improved or resolved; oral/nasogastric intake tolerated, without any absorption issues; absence of indication for prolonged intravenous therapy (i.e. meningitis, febrile neutropenia, bloodstream infection, endocarditis, osteomyelitis, etc.); suitable oral agent available; and likelihood that patient will be concordant with oral therapy (i.e. paediatric patients who may dislike the taste and consistency of oral therapy) (Kuper, 2008; Thompson *et al.*, 2015; Gasparetto *et al.*, 2019). Each one of these criteria requires nurses to demonstrate clinical expertise and advocacy for their patients, and for them to communicate promptly with other professionals (e.g. following up laboratory results, or discussing the clinical improvement of patients during ward rounds).

There are multiple benefits associated with the prompt switch from intravenous to oral therapy among suitable patients (Bassetti *et al.*, 2018; Powell and Wilcock, 2019). For example, oral therapy incurs fewer treatment costs, requires decreased nursing time in terms of preparation and administration, contributes to a reduced risk of iatrogenias such as vascular catheter-associated infection, and extravasations. Furthermore, this switch can reduce length of hospital stay, which leads to patient satisfaction (Cyriac *et al.*, 2014).

Nurses should be familiar with the three main approaches to intravenous to oral switch, namely sequential therapy, switch therapy and de-escalation therapy (PHE, 2015). In a sequential therapy, the prescriber would maintain the same antimicrobial but would convert it from intravenous to oral form (i.e. linezolid) (Al-Hasan *et al.*, 2019). In a switch, the intravenous antimicrobial is replaced with another antimicrobial in oral form but with similar potency. Finally, de-escalation of therapy takes place when an intravenous antimicrobial is replaced with an oral antimicrobial of reduced potency, narrower spectrum, tailored to the pathogen or different frequency, and frequently following laboratory confirmation of infection, micro-organism and susceptibility.

Despite the benefits for patients, nurses and healthcare organizations, the engagement of nurses with intravenous to oral switch has been insufficient, with some evidence already available about the determinants of such suboptimal participation (Ha *et al.*, 2019). For instance, the vast majority of patients eligible to switch (~60–75%) are not offered the option (Hermsen *et al.*, 2013). Fisher *et al.* (2018) acknowledged the barriers and facilitators of nurse-driven AMS, and, in particular, the factors influencing the promotion of intravenous to oral antimicrobial switch (Table 5.2). Among these determinants, insufficient knowledge, lack of prompts and standardized procedures to establish the suitability of patients for the switch were identified.

Other authors have reported on interventions aiming to facilitate intravenous to oral switch initiated or supported by nurses as well as their wider participation in AMS. For example, Sumner *et al.* (2017) proposed the use of standardized 'scripts' (i.e. conversational interactions) (Box 5.2) and role-playing at crucial points along the decision-making process about antibiotic management, including when patients may be suitable for intravenous to oral switch.

Table 5.2. Barriers and facilitators to nurse-driven AMS – focus on intravenous to oral antimicrobial switch.

Barrier	Facilitator
Lack of prompts reminding nurses to assess suitability of patients for intravenous to oral switch	Ability to assess patients for appropriateness of intravenous to oral switch
Absence of standardized pathway to determine suitability of patients for intravenous to oral switch	Capable of communicating results of intravenous to oral switch assessment to multidisciplinary team
Lack of cooperation and support from prescriber	Able to actively participate in multidisciplinary activities (i.e. ward rounds) to discuss intravenous to oral switch
Lack of prescriber accessibility	Collaboration with nursing colleagues to confirm patients' appropriateness for intravenous to oral switch
Lack of self-confidence	Knowledge about patients' clinical progress and improvement
Intravenous to oral switch perceived as independent or sole role of prescriber	Availability of education and training resources on intravenous to oral switch
Concerns with adverse consequences of intravenous to oral antimicrobial switch	Confidence in becoming leaders in promoting IV to PO step-down with the support of colleagues and other healthcare professionals
Intravenous to oral switch seen as a low priority activity for nurses	Increased nursing efficiency in administering and monitoring PO versus IV antimicrobials

Box 5.2. Proposed standardized conversational interaction on antibiotic therapy.

'The sensitivities on Mrs Jones culture(s) have been received from the laboratory. The report indicates Mrs Jones is sensitive/resistant to [ANTIBIOTIC]. She is currently receiving the following antibiotic(s): [ANTIBIOTIC]. Do you want to continue this/these antibiotic(s)?'

Improving and refining messages referring to antimicrobial management could help nurses embrace stewardship roles and, in particular, engage in intravenous to oral switch decisions. For example, Grayson *et al.* (2015), in their tailoring of activities to professional personality profiles, recommended that nurses focused on the benefit to patients rather than the pharmacological appropriateness of the prescription. Equally, Raybardhan *et al.* (2017) achieved a substantial improvement in the number of patients for whom antibiotics were discussed during ward rounds, including suitability for intravenous to oral switch, following a multidisciplinary quality improvement intervention.

The Appropriateness of Antimicrobial Administration Models such as Outpatient Parenteral Antimicrobial Therapy (OPAT)

Outpatient parenteral antimicrobial therapy (OPAT) is another option to optimize antimicrobial treatment for patients where nurses can have a crucial role. Essentially, OPAT involves the administration of at least two doses of intravenous antimicrobial therapy on different days without admission to the hospital (Norris *et al.*, 2019). Thus, the goal of OPAT programmes is to enable patients to complete antimicrobial therapy safely and effectively, in their home or a suitable outpatient department. Completing treatment at home would avoid patients being exposed to hospital-associated pathogens and subsequent infections; this would naturally decrease their length of hospital stay (and even circumvent the need for such admission) and reduce the costs of healthcare and treatment (Ross, 2010; Lane *et al.*, 2014). Other patient-centred advantages of OPAT include psychological and social benefits, by allowing people to be at home, leading to more comfortable recovery.

The systematic review by Minton *et al.* in 2017, funded by the National Institute for Health Research in the UK, made these benefits explicit, with some caveats. Whilst there was no impact on the total duration of intravenous treatment, when compared with in-patient antibiotic courses, the effect of the OPAT pathway on the cure rate appeared to suggest that OPAT led to higher results compared with in-patient treatments. However, such superiority disappeared when studies reporting on multiple or unspecified OPAT treatment models were removed and only specific models were individually compared. In such instance, outpatient attendance appeared to have a lower rate of cure, OPAT by a specialist nurse had a higher success rate and, finally, OPAT by a general nurse had no impact. Despite these differences, the review indicated that there seemed to be no negative impact of OPAT on drug-related side effects or serious adverse events such as deaths when compared with in-patient-delivered treatments. Finally, OPAT was modelled to be more cost-effective than in-patient therapy in several instances, and achieved a rotund success in terms of patient satisfaction, i.e. only 5% of home OPAT patients would have preferred hospital treatment, whilst 35% of those treated at hospital would have preferred home treatment.

Such views have been corroborated subsequently by Berrevoets *et al.* (2018), who explored elements of patient-centredness afforded by OPAT services. Perhaps not surprisingly, the participants in this study, conducted in The Netherlands, identified the relative freedom allowed by the treatment (i.e. being at home and being able to negotiate the therapy with their carers, etc.) as one of its most valuable benefits.

In order for patients to be eligible for OPAT, they must be adequately assessed, ensuring that the clinical care required does not demand hospital admission, the condition of the patient is stable and improving, the outpatient or home environment are supportive and conducive to recovery and allow for monitoring of therapeutic progress, the patient or carer is skilled and capable to administer the medication safely, and they are willing and able to participate in the treatment. All these assessments and safeguards require optimal practice of essential nursing skills, including patient education and communication, evaluation of health literacy deficits, as well as facilitating shared decision-making and care partnership.

Use of Perioperative Prophylactic Antimicrobials to Prevent Surgical Site Infection

Surgical site infections (SSIs) are defined as infections that occur up to 30 days following surgery (or up to one year following surgery in patients who receive implants) and affect either the incision or deep tissue at the operation site (Owens and Stoessel, 2008). SSIs are the most surveyed and frequent healthcare-associated infections (HAIs) in low- and middle-income countries (Allegranzi *et al.*, 2018), and the second most frequent SSIs in Europe and USA (Magill *et al.*, 2014), accounting for approximately 20% of all HAIs among hospitalized patients in Europe and ~38% of all post-operative complications (ECDC, 2013). The consequences of such infections can also be dramatic, with patients experiencing an SSI 2–11 times more likely to die, compared to those persons who were operated on and did not develop such infections (Kirkland *et al.*, 1999 Engemann *et al.*, 2003). Multiple factors related to patients or the surgical procedure can influence the risk of developing SSIs. These risk factors include functional status, obesity, complicated or emergency surgery, prolonged surgical duration, respiratory conditions such as chronic obstructive pulmonary disease, diabetes, smoking, peripheral vascular disease or limb ischemia, hypertension, coronary artery disease, coagulation bleeding disorders, renal disease, preoperative sepsis and gender (Wiseman *et al.*, 2015).

As with many other related infection prevention and control and antimicrobial stewardship interventions, the participation of nurses in quality improvement interventions aiming to prevent surgical site infections demands the provision of technically excellent, patient-centred care which includes not only attention to the patient and surgery factors mentioned before, but also a high degree of situational awareness in the operating theatre and appropriate levels of communication between nurses and other members of the multidisciplinary team.

Reviewing the WHO Global Guidelines for the Prevention of Surgical Site Infection (2016), for example, demonstrates the ownership and responsibility towards prevention of SSIs that nurses can, and must, have. Many of the recommended steps and interventions are led by nurses or fall within essential areas of nursing competence. Such jurisdiction also extends to those recommendations focused on administration of antimicrobials; for example, the decolonization of persons with MRSA with mupirocin ointment, when required, or the timely administration of antimicrobial prophylaxis.

Such prophylactic use of antimicrobials refers to administration prior to any occurrence of contamination, so the release and dissemination of the antimicrobial to the surgical site ensures sufficient tissue concentration of the antibiotic in the target area. This way, the growth of any contaminating micro-organisms would be inhibited, thus reducing the risk of infection and in some cases (e.g. surgical implants) the need to repeat or reverse the surgery. The appropriate selection of prophylactic antibiotics and the timing of administration are thus crucial to reducing the likelihood of SSIs. Guidance regarding these two factors is based upon the type of surgery (i.e. whether the surgical wound is clean, clean-contaminated or contaminated) (Mulder *et al.*, 2019; Wainberg et al., 2019). In addition to ensuring that antimicrobials are administered on time, nurses should also contribute to monitoring that prophylactic doses do not continue inappropriately following surgery, which would result, effectively, in patients receiving therapeutic rather than prophylactic courses of antibiotics.

In terms of communication and interprofessional practice, Troughton *et al.* (2019) interviewed healthcare professionals involved in the surgical pathway to describe any determinants of infection control in surgery, and, in particular, to highlight any social and contextual elements. Their analysis identified shared ownership of surgical site infections as well as team hierarchy as drivers of practice. Such findings have been replicated in other settings worldwide, stressing the impact that influential individuals, including nurses, could have to ensure motivation of teams, transmission of knowledge about optimal antibiotic practice, multidisciplinary engagement, active leadership and timely feedback of infection rates (Clack *et al.*, 2019; Nasiri *et al.*, 2019).

Factors that Can Influence Antimicrobial Prescribing and the Implications for Antimicrobial Stewardship Programmes

As this chapter has highlighted, optimal antimicrobial prescribing and management requires much more than sufficient technical knowledge (i.e. about pharmacology, pathology or therapeutics), and must be coupled with an awareness of the multiple organizational, behavioural and cultural factors that can end up influencing the prescription of an antimicrobial, the length of such prescription or the discussion to review a therapy. It is clear that antimicrobial prescribing is influenced by multiple factors, both in the community and hospitals, and therefore nurses should be attuned to the effect that such factors can have in their individual practice and as part of interdisciplinary teams jointly managing antibiotics.

Essentially, factors influencing antimicrobial use can relate to the prescriber, the patient and the organizational environment and climate (Tromp *et al.*, 2010). Psychological biases, such as fear of failure or perceptions about likelihood of poor clinical outcomes or clinical effectiveness, have been reported as major factors influencing prescribers (Touboul-Lundgren *et al.*, 2015). Other factors that influence prescribers include uncertainty with regard to the working diagnosis (Teixeira Rodrigues *et al.*, 2013), prognosis (Broom *et al.*, 2014), or a desire to maintain clinician–patient relationship (Livorsi *et al.*, 2015). Equally, perceptions and unspoken assumptions by prescribers about the desire of patients to receive an antibiotic, for themselves or for their child, for example, strongly predict whether patients will indeed receive such prescription (McKay *et al.*, 2016). Further, lack of diagnostic tests (and particularly rapid tests), commercial influences on prescribing, high workload or limited time per patient influence antimicrobial prescribing practice and may not be completely acknowledged (Moragas *et al.*, 2019).

Sociocultural factors and factors related to interprofessional dynamics also play an important role in the prescription of antimicrobials. For example, whilst it is very likely that institutions will have guidelines and protocols about preferred and optimal antibiotic use, as seen in this chapter, it is also likely that prescribers will be influenced in their practice not just by the indications in the documents but by senior and more experienced clinicians who may have their own preferences for antimicrobials and would have their own opinion about the recommendations included in the policies. Ultimately, the influence of these senior figures and the associated hierarchy, coupled with unspoken rules implicitly known and understood by all professional groups involved in the management of antimicrobials, including nurses, is so powerful that it discourages any modification or discussion of suboptimal practice not aligned with institutional preferences (Charani *et al.*, 2013). These contextual and organizational factors appear to be so powerful that some authors have suggested that it would not be possible to be involved with, and improve, prescribing practice without taking them into account (Donisi *et al.*, 2019).

The multiple, intertwined factors that influence optimal prescribing are reflected in the variety of quality improvement interventions already reported upon. For example, the Cochrane review by Davey *et al.* (2017) identified more than 200 studies, conducted worldwide, of various designs, concluding that enablement strategies improved the effect of interventions, and that feedback further increased such effect. These findings would be important for nurses engaged in antimicrobial management, as they may wish to reflect upon how best to provide such feedback, as well as consider how to deal with the organizational and team influences.

Antimicrobial Stewardship in Action

Case study 1

The Safety and Quality Improvement Department at your hospital recently identified that the performance of your ward related to intravenous to oral antibiotic switches was below the average of the rest of your organization. The manager in your ward has organized a meeting with the nurses to think about how to increase the participation of nurses in antibiotic switches.

Continued

Case study 1 Continued.

Question: What suggestions could you offer at this team meeting?

Answer: There are several interventions focused on nurses that could be implemented. Before that, though, it may be useful to identify if the patient mix in the ward is very different to that of the other wards, which may explain differences in decisions about medication switches (e.g. if most patients on the ward had health problems that made oral medication intake difficult).

You could volunteer to participate or lead, together with your medical and pharmacy colleagues, in the evaluation of the existing intravenous to oral antibiotic switch pathway, to identify if it is clear and precise, including where nurses could input on the decision-making process about medications. It may be that there is no pathway at all. You could offer to develop a training session so that all nurses in the team can proficiently evaluate if patients are suitable for the switch. Additionally, the training may require developing interventions so that nurses communicate with other professionals about the suitability of patients.

Also, you could suggest that the staffing rota is organized with the timing of ward rounds in mind, so enough nurses are available to attend the rounds and that therefore up-to-date information about patients, including their clinical improvement and tolerance of oral medicines and fluids, can be taken into account. Such attendance may require a revision of the information documented on nursing notes about patients' progress as well.

You could suggest that the electronic health record includes prompts at crucial times in the patient's pathway (i.e. 24 h or 48–72 h). If you use paper charts, these may be redesigned in collaboration with pharmacy colleagues so that switch reminders or prompts are included.

You could offer to develop a standard operating procedure or protocol for the review of availability of requested, pending and received microbiology laboratory results, and how best to communicate them to prescribers. Perhaps these results (or lack of them) could also be included in the information that is systematically offered during ward rounds, which would inform decisions to narrow the antimicrobial therapy.

Case study 2

Mr A is a 44-year-old postman who was bitten on the hand by a neighbour's dog 2 days previously. He attended the accident and emergency department of the local hospital at the end of his shift. Following assessment at the hospital, a course of intravenous antibiotics was prescribed and Mr A was admitted to the ward. Whilst on the ward, Mr A overhears one of the nurses talking to another patient, also on intravenous antibiotics, who is being discharged and sent to the OPAT. Mr A would also like to know more about that OPAT option as he would really like to be back home with his family as soon as possible.

Question: What information would you give to Mr A?

Answer: You would mention to Mr A that OPAT means that antibiotics are given to people as outpatients (i.e. not in hospital). However, this does not mean that

Continued

> **Case study 2** Continued.
>
> patients will not need to come to the hospital every day so that the antibiotic can be given.
>
> You would also mention that, depending on the infection being treated, the features of the patient and the availability of local clinical services, the treatment could be given at home by community nurses, or Mr A or his spouse could even learn how to prepare the antibiotic and self-administer it.
>
> You would say that before being able to use the OPAT service, there should be clear improvement in Mr A's infected wound and that, depending on the antibiotic use, the clinical team would ensure that Mr A understands well that side effects and complications require review by the team and/or attendance at accident and emergency or review by the local family doctor or nurse.

Key Points

- Nurses can play a crucial role in the prompt identification of patients with sepsis, ensuring that treatment interventions are implemented and actioned without delay.
- National, organizational or local antibiotic guidelines can be beneficial, and nurses must proactively participate in guideline writing and reviewing whilst being cognizant of the shortcomings these documents can have.
- Supporting patients' concordance with antibiotic therapy is an essential nursing activity. The evidence regarding optimal duration of antibiotic courses is quite dynamic, demanding that nurses remain engaged with studies focused on treatment duration, clinical effectiveness of shorter courses and reported disadvantages.
- In order to provide patient-centred care, nurses should evaluate the therapeutic options available to patients, including outpatient antibiotic therapy or intravenous to oral switch, which offer benefits for patients as well as organizations.
- The participation of nurses in quality improvement interventions, either focused on antimicrobial prescribing or the prevention of surgical site infections, demands attention to the patient, the clinical factors and other sociocultural determinants of excellent practice.

Further Reading

Goff, D.A., Kullar, R., Goldstein, E.J.C., Gilchrist, M., Nathwani, D., *et al.* (2017) A global call from five countries to collaborate in antibiotic stewardship: united we succeed, divided we might fail. *The Lancet Infectious Dieases* 17(2), e56–e63. DOI: 10.1016/S1473-3099(16)30386-3

Hatcher, J., Costelloe, C., Cele, R., Viljanen, A., Samarasinghe, D., *et al.* Factors associated with successful completion of outpatient parenteral antibiotic therapy (OPAT): a 10-year review from a large West London service. *International Journal of Antimicrobial Agents* 54(2), 207–214. DOI: 10.1016/j. ijantimicag.2019.04.008

Wieringa, S. and Greenhalgh, T. (2015) 10 years of mindlines: a systematic review and commentary. *Implementation Science* 10(45). Available at: https://www.ncbi.nlm.nih.gov/pubmed/25890280 (accessed 21 September 2019).

References

Al-Hasan, M.N. and Rac, H. (2019) Transition from intravenous to oral antimicrobial therapy in patients with uncomplicated and complicated bloodstream infections. *Clinical Microbiology and Infection*. Available at: https://doi.org/10.1016/j.cmi.2019.05.012 (accessed 21 September 2019).

Allegranzi, B., Aiken, A.M., Zeynep Kubilay, N., Nthumba, P., Barasa, J., *et al*. (2018) A multimodal infection control and patient safety intervention to reduce surgical site infections in Africa: a multicentre, before-after, cohort study. *The Lancet Infectious Diseases* 18(5), 507–515. DOI: 10.1016/S1473-3099(18)30107-5

Bassetti, M., Eckmann, C., Peghin, M., Carnelutti, A. and Righi, E. (2018) When to switch to an oral treatment and/or to discharge a patient with skin and soft tissue infections. *Current Opinion in Infectious Diseases* 31(2),163–169. DOI: 10.1097/QCO.0000000000000434

Berrevoets, M.A.H., Oerlemans, A.J.M., Tromp, M., Kullberg, B.J., Ten Oever, J., *et al*. (2018) Quality of outpatient parenteral antimicrobial therapy (OPAT) care from the patient's perspective: a qualitative study. *BMJ Open* 8(11), e024564. DOI: 10.1136/bmjopen-2018-024564

Broom, A., Broom, J. and Kirby, E. (2014) Cultures of resistance? A Bourdieusian analysis of doctors' antibiotic prescribing. *Social Science and Medicine* 110, 81–88.

Broom, A., Broom, J., Kirby, E. and Scambler, G. (2017) Nurses as antibiotic brokers: institutionalized praxis in the hospital. *Qualitative Health Research* 27(13), 1924–1935. DOI: 10.1177/1049732316679953

Chambers, A., MacFarlane, S., Zvonar, R., Evans, G., Moore, J., *et al*. (2019) A recipe for antimicrobial stewardship success: using intervention mapping to develop a program to reduce antibiotic overuse in long-term care. *Infection Control & Hospital Epidemiology* 40(1), 24–31. DOI: 10.1017/ice.2018.281

Charani, E., Castro-Sánchez, E., Sevdalis, N., Kyratsis, Y., Drumright, L., *et al*. (2013) Understanding the determinants of antimicrobial prescribing within hospitals: the role of 'prescribing etiquette'. *Clinical Infectious Diseases* 57(2), 188–196. DOI: 10.1093/cid/cit212

Chiotos, K., Tamma, P. and Gerber, J. (2019) Antibiotic stewardship in the intensive care unit: challenges and opportunities. *Infection Control & Hospital Epidemiology* 40(6), 693–698.

Clack, L., Willi, U., Berenholtz, S., Aiken, A.M., Allegranzi, B. and Sax, H. (2019) Implementation of a surgical unit-based safety programme in African hospitals: a multicentre qualitative study. *Antimicrobial Resistance Infection Control* 8, art. 91. DOI: 10.1186/s13756-019-0541-3

Coates, E., Villarreal, A., Gordanier, C. and Pomernacki, L. (2015) Sepsis power hour: a nursing driven protocol improves timeliness of sepsis care. *Journal of Hospital Medicine* 10 (suppl 2).

Costelloe, C., Metcalfe, C., Lovering, A., Mant, D. and Hay, A.D. (2010) Effect of antibiotic prescribing in primary care on antimicrobial resistance in individual patients: systematic review and meta-analysis. *BMJ* 340, c2096. DOI: 10.1136/bmj.c2096 pmid:20483949

Cyriac, J.M. and James, E. (2014) Switch over from intravenous to oral therapy: a concise overview. *Journal of Pharmacology and Pharmacotherapeutics* 5(2), 83–87.

Davey, P., Marwick, C.A., Scott, C.L., Charani, E., McNeil, K., Brown, E., *et al.* (2017) Interventions to improve antibiotic prescribing practices for hospital inpatients. *Cochrane Database of Systematic Reviews* 2, CD003543. DOI: 10.1002/14651858.CD003543.pub4

Dawson-Hahn, E.E., Mickan, S., Onakpoya, I., Roberts, N., Kronman, M., *et al.* (2017) Short-course versus long-course oral antibiotic treatment for infections treated in outpatient settings: a review of systematic reviews. *Family Practice* 34, 511–519. Available at: https://www.ncbi.nlm.nih.gov/pmc/articles/PMC6390420/ (accessed 21 September 2019).

Donisi, V., Sibani, M., Carrara, E., Del Piccolo, L., Rimondini, M., *et al.* (2019) Emotional, cognitive and social factors of antimicrobial prescribing: can antimicrobial stewardship intervention be effective without addressing psycho-social factors? *Journal of Antimicrobial Chemotherapy* 74(10), 2844–2847. Available at: https://doi.org/10.1093/jac/dkz308 (accessed 21 September 2019).

ECDC (European Centre for Disease Prevention and Control) (2013) Point prevalence survey of healthcare-associated infections and antimicrobial use in European acute care hospitals 2011–2012. Available at: http://ecdc.europa.eu/en/publications/Publications/healthcare-associated-infections-antimicrobial-use-PPS.pdf (accessed 21 September 2019).

Engemann, J.J., Carmeli, Y., Cosgrove, S.E., Fowler, V.G., Bronstein, M.Z., *et al.* Adverse clinical and economic outcomes attributable to methicillin resistance among patients with Staphylococcus aureus surgical site infection. *Clinical Infectious Diseases* 36, 592–598.

Fisher, C.C., Cox, V.C., Gorman, S.K., Lesko, N., Holdsworth, K., *et al.* (2018) A theory-informed assessment of the barriers and facilitators to nurse-driven antimicrobial stewardship. *American Journal of Infection Control* 6(12),1365–1369.

Fitzpatrick, F., Tarrant, C., Hamilton, V., Kiernan, F.M., Jenkins, D. and Krockow, E.M. (2019) Sepsis and antimicrobial stewardship: two sides of the same coin. *BMJ Quality & Safety*. DOI: 10.1136/bmjqs-2019-009445

Fleischmann, C., Scherag, A., Adhikari, N.K., Hartog, C.S., Tsaqanos, T., *et al.* (2016) Assessment of global incidence and mortality of hospital-treated sepsis: current estimates and limitations. *American Journal of Respiratory Critical Care Medicine* 193(3), 259–272.

Gasparetto, J., Tuon, F.F., dos Santos Oliveira, D., Zequinao, T., Pipolo, G.R., *et al.* (2019) Intravenous-to-oral antibiotic switch therapy: a cross-sectional study in critical care units. *BMC Infectious Diseases* 19, 650.

Grayson, M.L., Macesic, N., Huang, G.K., Bond, K., Fletcher, J., *et al.* (2015) Use of an innovative personality-mindset profiling tool to guide culture-change strategies among different healthcare worker groups. *PLOS One* 10(10), e0140509.

Greenhalgh, T., Howick, J. and Maskrey, N. (2014) Evidence-based medicine: a movement in crisis? *BMJ* 348, g3725.

Ha, D.R., Forte, M.B., Olans, R.D., OYong, K., Olans, R.N., *et al.* (2019) A multidisciplinary approach to incorporate bedside nurses into antimicrobial stewardship and infection prevention. *Joint Commission Journal on Quality and Patient Safety/Joint Commission Resources* 45(9). DOI: 10.1016/j.jcjq.2019.03.003

Hermsen, E.D., Shull, S.S., Richter, L.M. and Qiu, F. (2013) Failure of a pharmacist-initiated antimicrobial step-down protocol to impact physician prescribing behavior or patient outcomes: a quasi-experimental cross-over study. *Journal of Hospital Administration* 2(4), 63–70.

Kirkland, K.B., Briggs, J.P., Trivette, S.L., Wilkinson, W.E. and Sexton, D.J. (1999) The impact of surgical-site infections in the 1990s: attributable mortality, excess length of hospitalization, and extra costs. *Infection Control & Hospital Epidemiology* 20, 725–730.

Kozyrskyj, A.L., Klassen, T.P., Moffatt, M. and Harvey, K. (2010) Short-course antibiotics for acute otitis media. *Cochrane Database of Systematic Reviews*. DOI: 10.1002/14651858.CD001095.pub2

Kumar, P., Jordan, M., Caesar, J. and Miller, S. (2015) Improving the management of sepsis in a district general hospital by implementing the Sepsis Six recommendations. *BMJ Quality Improvement Programme*. Available at: https://bmjopenquality.bmj.com/content/4/1/u207871.w4032 (accessed 21 September 2019).

Kuper, K.M. (2008) Intravenous to oral therapy conversion. In: *Competence Assessment Tools*. NIPEC, Belfast, UK.

Lane, M.A., Marschall, J., Beekmann, S.E., Polgreen, P.M., Banerjee, R., Hersh, A.L. and Babcock, H.M. (2014) Outpatient parenteral antimicrobial therapy practices among adult infectious disease physicians. *Infection Control & Hospital Epidemiology* 35(7), 839–844.

Levy, M.M., Evans, L.E. and Rhodes, A. (2018) The surviving sepsis campaign bundle. *Critical Care Medicine* 45(3), 486–552.

Livorsi, D., Comer, A., Matthias, M.S., Perencevich, E.N. and Bair, M.J. (2015) Factors influencing antibiotic-prescribing decisions among inpatient physicians: a qualitative investigation. *Infection Control & Hospital Epidemiology* 36, 1065–1072.

Llewelyn, M.J., Fitzpatrick, J.M., Darwin, E., Tonkin-Crine, S., Cliff, G., *et al.* (2017) The antibiotic course has had its day. *BMJ* 358, 3418.

Magill, S.S., Edwards, J.R., Bamberg, W., Beldavs, Z.G., Dumyati, G., Kainer, M.A., *et al.* (2014) Multistate point-prevalence survey of health care-associated infections. *New England Journal of Medicine* 370, 1198–1208.

McKay, R., Mah, A., Law, M.R., McGrail, K. and Patrick, D.M. (2016) Systematic review of factors associated with antibiotic prescribing for respiratory tract infections. *Antimicrobial Agents and Chemotherapy* 60(7), 4106–4118. DOI: 10.1128/AAC.00209-16

Milo, G., Katchman, E.A., Paul, M., Christiaens, T., Baerheim, A. and Leibovici, L. (2005) Duration of antibacterial treatment for uncomplicated urinary tract infection in women. *Cochrane Database of Systematic Reviews*. DOI: 10.1002/14651858.CD004682.pub2

Minton, J., Czoski Murray, C., Meads, D., Hess, S., Varqas-Palacios, A., *et al.* (2017) The Community IntraVenous Antibiotic Study (CIVAS): a mixed-methods evaluation of patient preferences for and cost-effectiveness of different service models for delivering outpatient parenteral antimicrobial therapy. *Health Services and Delivery Research* 5(6).

Moore, L.S.P., Freeman, R., Gilchrist, M.J., Gharbi, M., Brannigan, E., *et al.* (2014) Homogeneity of antimicrobial policy, yet heterogeneity of antimicrobial resistance: antimicrobial non-susceptibility among 108717 clinical isolates from primary, secondary and tertiary care patients in London. *Journal of Antimicrobial Chemotherapy* 69(12), 3409–3422.

Moragas, A., Molero, J.M., Bjerrum, L. and Llor, C. (2019) General practitioners' opinions and perceptions about antibiotic use for respiratory tract infections in primary care. *Atención Primaria* 51(7), 460–461. Available at: https://doi.org/10.1016/j.aprim.2019.01.011 (accessed 21 September 2019).

Mulder, T., Crolla, R., Kluytmans-van den Bergh, M., van Mourik, M., Romme, J., *et al.* (2019) Preoperative oral antibiotic prophylaxis reduces surgical site infections after elective colorectal surgery: results from a before–after study. *Clinical Infectious Diseases* 69(1), 93–99.

Nasiri, A., Balouchi, A., Rezaie-Keikhaie, K., Bouya, S., Sheyback, M. and Rawajfah, O.A. (2019) Knowledge, attitude, practice, and clinical recommendation toward infection control and prevention standards among nurses: a systematic review. *American Journal of Infection Control* 47(7), 827–833. DOI: 10.1016/j.ajic.2018.11.022

NICE (2015) Antimicrobial stewardship: systems and processes for effective antimicrobial medicine use. Available at: https://www.nice.org.uk/guidance/ng15 (accessed 22 September 2019).

NICE (2016) Sepsis: recognition, diagnosis and early management. Available at: https://www.nice.org.uk/guidance/ng51 (accessed 22 September 2019).

Norris, A.H., Shrestha, N.K., Allison, G.M. *et al.* (2018) Infectious Diseases Society of America clinical practice guideline for the management of outpatient parenteral antimicrobial therapy. *Clinical Infectious Diseases* 68:e1.

Owens, C.D. and Stoessel, K. (2008) Surgical site infections: epidemiology, microbiology and prevention. *Journal of Hospital Infection* 70(Suppl 2),3–10. DOI: 10.1016/S0195-6701(08)60017-1

Pouwels, K.B., Freeman, R., Muller-Pebody, B., Rooney, G., Henderson, K.L., *et al.* (2018) Association between use of different antibiotics and trimethoprim resistance: going beyond the obvious crude association. *Journal of Antimicrobial Chemotherapy* 73, 1700–1707. Available at: https://www.ncbi.nlm.nih.gov/pubmed/29394363 (accessed 22 September 2019).

Pouwels K.B., Hopkins, S., Llewelyn, M.J., Walker, A.S., McNulty, C.A.M., *et al.* (2019) Duration of antibiotic treatment for common infections in English primary care: cross sectional analysis and comparison with guidelines. *BMJ* 364, l440.

Powell, N. and Wilcock, M. (2019) Challenging the supremacy of intravenous antibiotics. *Drug and Therapeutics Bulletin* 57(1), 2.

Public Health England (2015) Antimicrobial Stewardship: *Start Smart – Then Focus. Antimicrobial Stewardship Toolkit for English Hospitals.* Available at: https://assets.publishing.service.gov.uk/government/uploads/system/uploads/attachment_data/file/417032/Start_Smart_Then_Focus_FINAL.PDF (accessed 20 December 2019).

Raybardhan, S., Chung, B., Ferreira, D., Bitton, M., Shin, P. *et al.* (2017) Nurse prompting for prescriber-led review of antimicrobial use in the critical care unit: a quality improvement intervention with controlled interrupted time series analysis. *Open Forum Infectious Diseases* 4(Suppl 1), S278.

Rhodes, A., Evans, L.E., Alhazzani, W., Levy, M.M., Antonelli, M., *et al.* (2016) Surviving sepsis campaign: international guidelines for management of sepsis and septic shock. *Intensive Care Medicine* 43(3), 304–377.

Ross Nolet, B. (2010) Update and overview of outpatient parenteral antimicrobial therapy regulations and reimbursement. *Clinical Infectious Diseases* 51(Suppl 2), S216–S219. DOI: 10.1086/653522

Rzewuska, M., Charani, E., Clarkson, J.E., Davey, P.G., Duncan, E.M., Francis, J.J. *et al.* (2019) Joint Programming Initiative on Antimicrobial Resistance (JPIAMR) Working Group on Behavioural Approaches to Antibiotic Stewardship Programs. Prioritizing research areas for antibiotic stewardship programmes in hospitals: a behavioural perspective consensus paper. *Clinical Microbiology and Infection* 25(2),163–168. DOI: 10.1016/j.cmi.2018.08.020

Singer, M., Deutschman, C.S., Seymour, C.W., *et al.* (2016) The third international consensus definitions for sepsis and septic shock (Sepsis-3). *JAMA* 315(8), 801–810.

Sumner, S., Forsyth, S., Collette-Merrill, K., Taylor, C., Vento, T., *et al.* (2017) Antibiotic stewardship: the role of clinical nurses and nurse educators. *Nurse Education Today* 60, 157–160.

Teixeira Rodrigues, A., Roque, F., Falcao, A., Figueiras, A. and Herdeiro, M.T. (2013) Understanding physician antibiotic prescribing behaviour: a systematic review of qualitative studies. *International Journal of Antimicrobial Agents* 41, 203–212.

Thompson, C., Zahradnik, M., Brown, A., Fleming, D.G. and Law, M. (2015) The use of an IV to PO clinical intervention form to improve antibiotic administration in a community based hospital. *BMJ Open Quality* 4(1). DOI: 10.1136/bmjquality.u200786.w2247

Touboul-Lundgren, P., Jensen, S., Drai, J. and Lindbæk, M. (2015) Identification of cultural determinants of antibiotic use cited in primary care in Europe: a mixed research synthesis study of integrated design 'Culture is all around us'. *BMC Public Health* 15, 908. DOI: 10.1186/s12889-015-2254-8

Troughton, R., Mariano, V., Campbell, A., Hettiaratchy, S., Holmes, A. and Birgand, G. (2019) Understanding determinants of infection control practices in surgery: the role of shared ownership and team hierarchy. *Antimicrobial Resistance and Infection Control* 8, art.116. DOI: 10.1186/s13756-019-0565-8

Tromp, M., Hulscher, M., Bleeker-Rovers, C.P., Peters, L., van den Berg, D.T., *et al.* (2010) The role of nurses in the recognition and treatment of patients with sepsis in the emergency department: a prospective before-and-after intervention study. *International Journal of Nursing Studies* 47, 1464–1473. DOI: 10.1016/j.ijnurstu.2010.04.007

Vincent, J.L., Moreno, R., Takala, J., Willatts, S., de Mendonca, A., *et al.* (1996) Working Group on Sepsis-Related Problems of the European Society of Intensive Care Medicine. The SOFA (Sepsis-related Organ Failure Assessment) score to describe organ dysfunction/failure. *Intensive Care Medicine* 22(7), 707–710.

Wainberg, S.K., Santos, N.C.L., Gabriel, F.C., de Vasconcelos, P., Nascimento, J.S., *et al.* (2019) Clinical practice guidelines for surgical antimicrobial prophylaxis: qualitative appraisals and synthesis of recommendations. *Journal of Evaluation in Clinical Practice* 25, 591–602. Available at: https://doi.org/10.1111/jep.12992 (accessed 22 September 2019).

Wilson, H.L., Daveson, K. and Del Mar, C.B. (2019) Optimal antimicrobial duration for common bacterial infections. *Australian Prescriber* 42(1), 5–9. DOI:10.18773/austprescr.2019.001

Wiseman, J.T., Fernandez-Taylor, S., Barnes, M., Saunders, R.S., Saha, S., *et al.* (2015) Predictors of surgical site infection after hospital discharge in patients undergoing major vascular surgery. *Journal of Vascular Surgery* 62(4), 1023–1031.

WHO (World Health Organization) (2016) *Global Guidelines for the Prevention of Surgical Site Infection*. WHO, Geneva, Switzerland.

Person-centred Care   6

Emma Burnett[1,*] and Valerie Ness[2]

[1]*Associate Dean International & Academic Regional Lead for Middle East and North Africa, University of Dundee, UK;* [2]*Lecturer, School of Health and Life Sciences, Glasgow Caledonian University, UK*

Objective: For the student to integrate and value input and engagement of the patient/carer in designing and implementing care.

Introduction

Person-centred care (PCC) involves treating the person as an individual. It is a partnership between the healthcare professional and the patient, and, if appropriate, their family and carers, which is built on mutual trust and understanding (McCormack and McCance, 2016). The benefits of involving the patient in assessing and planning their own care include improved adherence to treatment, which is important in antimicrobial stewardship (AMS). This chapter will explore how using the key concepts of PCC, such as sharing information with patients, involving them in shared decision-making and listening to their needs, can improve healthcare professionals' AMS practice. The chapter will then discuss patients' expectations for an antimicrobial and management strategy embedded in PCC that can be adopted to ensure appropriate prescribing.

Patients/Carers as Integral Partners in Care – Sharing Information and Decision-making

The importance of patients and carers as an integral part of their own healthcare is a globally recognized, fundamental component of patient-centred care. Evidence demonstrates that when patients and carers are actively involved in their own care and in the decisions made about their care, it leads to better patient adherence to treatment, improved treatment outcomes and an enhanced patient experience (Ham

---

*Corresponding author: E.Burnett@dundee.ac.uk

et al., 2018). However, active patient involvement in their care does not consistently occur. A qualitative exploratory study was undertaken by Rawson *et al.* (2016) to understand patient engagement with decision-making for infection management in secondary care. Participants said they had 'completely lost ownership of their condition [infection]' and were rarely informed about the name of the antibiotic, duration of treatment and when the treatment would cease. All participants reported that they felt uninvolved in the decision-making process. These findings were replicated in another qualitative study exploring antibiotic decision-making within a surgical team (Charani *et al.*, 2017). This study found that at no time was the patient or carer involvement discussed. Despite recognition that patient and carer involvement is crucial, such evidence tells us that this is still not fully embedded into practice.

Shared decision-making involves the healthcare team, patients and carers working together to make the best decisions about treatment and care. To ensure that such collaboration happens effectively, information needs to be exchanged among all those involved, ensuring that all perspectives are voiced and equally considered. For example, healthcare professionals can provide evidence-based information about the infection, illness and symptoms, and therefore can communicate associated risks, benefits and potential outcomes of various care and treatment options. Patients and carers, on the other hand, can express their own understanding, perceptions, experiences, preferences, lifestyle and wishes. Butler *et al.* (2001) provides an example of key questions that healthcare professionals can use to ensure that patients are involved in decision-making (Box 6.1).

Using such an approach can provide the opportunity to elicit patients' and carers' values and preferences and target discussions about all possible options, enabling patient-centred joint decisions to be made. However, it is important to be mindful that many patients may be unfamiliar with playing an integral part in their care, and being a partner in decision-making. Additionally, it must be acknowledged that both healthcare professionals and patients can be influenced by cultural norms and mores regarding patient and professional roles. Taking the time to explain what is meant by shared decision-making can

---

Box 6.1. Shared decision-making prompt questions.

1. How would you feel if we talked things through and made a decision together about treatment?
2. How do you feel about antibiotics?
3. Would you like me to summarize the research evidence as it applies to your unique situation right now?
4. Would you like information about prevention and self-care, and if so, would you prefer to talk about it, take a leaflet or look at the internet site I could recommend?
5. How do you feel all of this applies to you?
6. Are you happy to talk about treatment and share the decision in their way?

prevent patients and their carers from feeling overwhelmed or anxious when asked such questions (Hoffman *et al.*, 2014).

Education and Support – the Provision of Information

For patients and carers to become integral partners in their care, it is crucial that they are provided with adequate and appropriate education and support to do so effectively. However, until recently most efforts have been largely focused on the education of healthcare professionals, with limited attention paid to patients and the public (Lee *et al.*, 2015). Research studies show that members of the public are not knowledgeable about antimicrobial stewardship, generally underestimate risks, are unaware of how their behaviour exposes them to microbial risks in their environment and do not believe that they can influence antimicrobial resistance (Brooks *et al.*, 2008; Harbarth *et al.*, 2015; Carter *et al.*, 2016).

Additionally, studies have established that patients and the public have insufficient knowledge and understanding of antimicrobial use, and therefore do not adhere to the advice provided to them by healthcare professionals. For example, many individuals do not finish their course of prescribed antibiotics; they save them for other occasions or share them among their friends and family (Pechère, 2001; Jose *et al.*, 2013; Vallin *et al.*, 2016; Jifar and Ayele, 2018; Tong *et al.*, 2018). Socio-economic factors, such as home address, occupation and family income, have been linked to such non-adherence to antimicrobial therapy.

To address these issues, according to Khan (2010), the prevailing approach to communicating about health-related risks is to simply flood people with as much 'perceived' robust information as possible on the assumption that it will be endorsed and acted upon. This is not an approach recommended, as different people would require different information for various reasons such as current perceptions, attitudes, knowledge and past experience. Additionally, Beagley (2011) points out several obstacles that can prevent effective delivery of healthcare information. These include health literacy, in that not every person is able to obtain, read, understand, appraise and apply information about their illness and how it relates to optimal antimicrobial use. Poor health literacy can also make it difficult for individuals to understand the information given to them to help them make informed decisions about their care (Panagioti *et al.*, 2018). It is therefore imperative that education and information resources are designed to accommodate people with a range of health literacy skills, and that interventions strive to address upstream determinants of health literacy (Sørensen *et al.*, 2013). Language and cultural differences between patients, carers and healthcare professionals may also cause misunderstandings and therefore impact on effective communication and learning (Almutairi, 2015). Cultural beliefs can influence people's perceptions of illness and healthcare; therefore, an understanding of these issues, without judgement, is vital so that patients can talk openly and honestly about their health issues. For example, some patients believe that tuberculosis is attributed to sharing eating utensils or even bewitchment (Viney *et al.*, 2014). This can lead to patients keeping their condition a secret for fear of being shunned.

Uncontrolled pain, anxiety and fear are also likely to impact on the willingness or ability of patients to communicate about health. For example, patients may not be able to articulate the level of pain they are in or feel comfortable saying that they are frightened about something (Ardalan *et al.*, 2018). Environmental aspects, such as a lack of time, can result in rushed consultations, and therefore the patient would not have adequate time to discuss matters of importance to them. Increased noise levels can lead to misunderstandings, and extreme temperature and lack of physical space can make patients feel physically uncomfortable, thereby reducing their concentration and ability to retain information (Stans *et al.*, 2016).

When ensuring patients and their carers receive person-centred education and support, learning styles must be considered carefully because people may have preferences to learn from different methods. For example, someone may be more of a visual learner, so pictures and images will be better to help them understand ideas and information rather than just text (Houts *et al.*, 2006). Individuals who may not be able to read are more likely to prefer audio information (Marcus, 2014), while others may prefer demonstration (Father and Stevens, 2008). For example, demonstrating hand-hygiene practices and different hand-hygiene activities may be much more effective than simply telling someone to wash their hands.

Listening to Patients – Respectful of the Expressed Needs of all Parties in Shaping and Delivering Care or Services

Patients and their carers approach healthcare professionals usually because they need help with something that is causing them problems. Healthcare professionals therefore need to understand exactly what the problem is and how it makes patients feel in order to respond to their needs (Scholl *et al.*, 2014). To respond effectively, healthcare professionals must establish trust and build rapport with patients. In doing so, professionals must take time to listen carefully and also to try to understand the patient's perspective.

However, good communication skills among healthcare professionals remain suboptimal. On average, doctors spend only 11 seconds listening to a patient before interrupting them, and interruptions occur in 67% of encounters (Ospina *et al.*, 2019). More worryingly, 37% of patients are interrupted before finishing what they want, or need, to say (Dyche and Swiderksi, 2005), which hinders how accurately clinicians can report patient problems and therefore plan therapeutic management.

However, while Mauksch (2017) agrees that listening to patients is a vital part of person-centred care, they argue that sometimes interruptions can indeed improve the quality of care and help the patient and healthcare professional make better use of their time together. An example they provide is when patients speak tangentially, bringing new and perhaps unconnected issues into the conversation, which can often take the interaction off course. This then results in a lengthened, often rushed, visit which may compromise quality of care. However, these researchers promote the use of respectfully phrased reassurance and recognition, combined with an interruption to redirect the patient to the main issue.

Berman and Chutka (2016) acknowledge that patients commonly complain that healthcare professionals do not listen to them, and so, to address this, they developed an assessment tool to teach medical students appropriate and effective communication skills. This comprehensive tool presents key elements of an effective healthcare professional–patient interaction, which includes giving a sincere introduction, maintaining appropriate eye contact throughout the discussion, leaning towards the patient from a comfortable distance, listening actively to the patient at all times, demonstrating sincere interest in the patient's emotional needs, keeping the patient focused throughout the interview and asking appropriate questions. The authors advocate that this tool will help professionals develop positive relationships with patients, which will foster high-quality, person-centred care.

To ensure that the patient's voice is heard, healthcare professionals must also ensure that patients understand the information provided to them in the way that it is intended. Further, clinicians must corroborate how patients have interpreted the information provided and, crucially, how they intend to act upon it. For example, have they fully understood the benefits and risks of the treatment or advice recommended, which will enable them to engage in decision-making? According to Ospina *et al.* (2019), this will allow patients and healthcare professionals to participate in meaningful conversations which will lay the foundation for patient-centered care.

Expectations for an Antimicrobial – Strategies to Deal with Expectations and the Need to Use Antimicrobials Appropriately

Why is involvement of the patient in information and decision-making so important for the success of AMS activities? One of the identified influences on inappropriate antibiotic prescribing is patient or carer expectation to receive an antibiotic, especially within primary care (Francis *et al.*, 2012; Mustafa *et al.*, 2014; Fletcher-Lartey *et al.*, 2016; Lum *et al.*, 2018). A recent review by King *et al.* (2018) also found agreement that clinicians' perceptions that patients want antibiotics influences them to prescribe inappropriately. These studies suggest that prescribers are influenced by a need to maintain or establish a therapeutic relationship with patients, a key component of person-centred care. However, such perspectives can present dilemmas, as a desire to benefit a given patient may need to be balanced with the wider public health and the potential collective harm caused by AMR. Indeed, a narrative review by Tarrant *et al.* (2019) found that the social dilemma of inappropriate antibiotic prescribing is complicated by individual actions and the outcome of AMR.

Patient expectations for antibiotics are also shaped by many factors. A study by Ebdell *et al.* (2013), which compared patients' expectations with data from a systematic review, found that some sociodemographic characteristics such as ethnic group, educational attainment and previous experience of receiving antibiotics, were associated with the likelihood of believing that antibiotics were always helpful. A more recent Australian study reported similar findings, with those who had never attended university, those speaking a language other than English, along with people younger than 65 years of age, more likely to expect or demand antibiotics for a cold or flu (Gaarslev *et al.*, 2016).

A mixed-methods study by Courtenay *et al.* (2017) explored the patient experience of non-medical prescriber (NMP) management of acute respiratory tract infections within primary care. One hundred and twenty patients were surveyed, 22 of which were followed up with interviews, and findings revealed that 43% of these patients expected to be prescribed an antibiotic. This was often the case where a patient was influenced by their history of receiving antibiotics for similar symptoms, or where, for an existing respiratory condition, a back-up prescription was the norm in case symptoms worsened. A qualitative study of nurse prescribers by Rowbotham *et al.* (2012) also found that prescribers found it challenging to manage patient expectations, often caused by patients' previous experience of receiving antibiotics, or due to a lack of patient understanding about the difference between a viral and bacterial infection. In a survey administered to patients, parents and caregivers (*n*=190) at seven primary care clinics and two urgent care locations in the USA, 53% of respondents thought that antibiotics worked well in treating viral infections. These same respondents were then twice as likely to expect a prescriber to give them an antibiotic when they had a cough or a common cold (Davis *et al.*, 2017).

There is evidence to suggest that in secondary care, patients are not engaged with their own treatment, reporting that they feel disempowered during episodes of infection (Rawson *et al.*, 2016). Similar to earlier discussions in this chapter, this was due to poor communication, with patients being told they had an infection and would receive an antibiotic but with very little other information. This meant that some of these patients sought information from other sources, such as online, where information is not necessarily reliable or person-centred. This influenced their future actions towards antibiotics and could lead to suboptimal therapeutic concordance. Not only could patients source information online but they could also purchase antibiotics. A cross-sectional analysis of online pharmacies selling to the UK public found that 16 (80%) of these pharmacies made decisions about antibiotic choice, dose and quantity initially driven by consumers, with nine not even requiring a prescription (Boyd *et al.*, 2017).

However, it is important to note that prescriber perception of patient desire for antibiotics may not be the same as the patient's desire. In a systematic review of factors associated with antibiotic prescribing for respiratory tract infections by McKay *et al.* (2016), it was found that although physicians' perceptions of patient desire for antibiotics were strongly associated with antibiotic prescribing, patient desire for antibiotics was only modestly associated with a prescription. In one of these studies (Coenen *et al.*, 2013), where patients explicitly asked for antibiotics, there was a trend towards reduced prescribing, thus suggesting that when patients are able to directly address the issue, a discussion can take place with the prescriber regarding the need for antibiotics.

Prescribers give two main reasons to justify perceptions of being under pressure to prescribe an antibiotic: time pressures and improved patient satisfaction. However, two studies found that the amount of time spent with a patient was not associated with antibiotic prescriptions (Linder *et al.*, 2003; Coco and Mainous, 2005); although a study of nurse and pharmacist prescribers by Courtenay *et al.* (2017) found that a lack of time restriction led to improved patient satisfaction, and was necessary to empower patients to take control of their illness, and therefore prevent unnecessary future consultations or expectations for antibiotics.

In relation to patient satisfaction, there is evidence to suggest that satisfaction does not influence antibiotic prescribing (Welschen *et al.*, 2004; Tonkin-Crine *et al.*, 2014). The study by Welschen *et al.* (2004) found that patients reported higher satisfaction with a visit to a doctor for an acute respiratory tract infection when they received information or reassurance rather than an antibiotic, and if patients were expecting an antibiotic, the odds of satisfaction with the consultation were similar among those who received information and those who received an antibiotic. In contrast, a mixed-methods study by Courtenay *et al.* (2017) found that there was a lower level of satisfaction with treatment among patients who expected, but did not receive, an antibiotic ($p<0.001$). However, the authors also identified that overall patient satisfaction with the entire consultation was not affected. Similarly, a recent retrospective observational study of adults with acute sinusitis by Sharp *et al.* (2017) assessed the association of antibiotic prescribing with patient-satisfaction scores. Among the 5169 consultations with medical prescribers, patient satisfaction was slightly higher (79.5% versus 75.4%) when antibiotics were prescribed. However, this still meant that 75.4% of consultations where antibiotics were not prescribed received a favourable satisfaction score. In this case other factors such as older patient age, more chronic conditions and doctor–patient relationship also positively influenced patient-satisfaction scores.

What strategies can prescribers therefore use to manage patient expectation? Within the non-medical prescriber (NMP) literature, there are positive findings to suggest that although prescribers feel pressure from patients, the majority do not give in to this pressure. The 34 prescribers in Rowbotham *et al.*'s (2012) study stated that they did not give in to this pressure and used strategies such as promoting self-management and educational resources to avoid prescribing an antibiotic. Courtenay's study also found that prescribers were not unduly influenced by patient expectations for an antibiotic and found that 'patient-centred' management strategies (including reassurance and providing information) were received by 86.7% of patients (Courtenay *et al.*, 2017). Since patients still report confusion about which illnesses can be managed by antibiotics, the development of easy-to-understand patient educational materials is necessary to address such incorrect beliefs. Discussing this material with the patient can help in empowering them to self-manage their illness.

Other factors that were found to influence patient satisfaction in Courtenay *et al.*'s (2017) study were: taking seriously any concerns from patients; conducting a physical examination; communicating the treatment plan and explaining the therapeutic decisions; and a lack of time restriction. Safety-netting in primary care settings (using delayed prescribing or follow-up appointments) was another management strategy that nurses in Courtenay *et al.*'s (2017) study used and again can be adopted by prescribers, especially in situations where patients are vulnerable or when gaining the results of a microbiological sample is appropriate prior to deciding on whether or not to commence antibiotics. The consideration of point-of-care testing in primary care for patients with suspected lower respiratory tract infections is recommended by the National Institute for Health and Care Excellence (NICE) in the UK in their guidance on pneumonia in adults (NICE, 2014).

There is further guidance from NICE (2015) that provides recommendations for prescribers and includes strategies that can be adopted to avoid prescribing. These include those mentioned above and again highlight that prescribers should take the time to discuss why prescribing an antibiotic may not be the best option, highlighting the harms and offering alternatives such as over-the-counter medication. Where an antibiotic is a treatment option, it is vital that prescribers document the reason in the patient's notes and prescribe the shortest effective course, the most appropriate dose and route of administration, and that this plan of care is fully discussed with the patient. Advanced communication skills are vital in managing patient expectations in conveying messages clearly and confidently and improving patient empowerment and self-management of conditions where antibiotics are not necessary. Where antibiotics are necessary, these same skills are required to educate patients in a person-centred way about how and why these important drugs are taken appropriately to protect them and society, and to preserve their future effectiveness.

Prescribers should be prepared and anticipate that patients may ask for, or expect, antibiotics. However, this should not influence their decision to prescribe them but rather influence the management strategies that they adopt to manage the situation. Prescribers should be made aware that not prescribing antibiotics will not reduce patient satisfaction, provided these strategies have been adopted. In thinking about appropriate strategies, it is also important to remember why patients have these expectations and what type of patients are more likely to expect them, as discussed previously in this chapter.

Antimicrobial Stewardship in Action

Shared decision-making (SDM) is a process where patients and healthcare professionals participate together in making decisions. Use of SDM may promote more appropriate use of antibiotics. A Cochrane systematic review found nine trials and one follow-up study that compared strategies used to facilitate the shared decision-making process in primary care. Results found that antibiotic prescribing reduced by almost 40% when these strategies were used compared with usual care (Coxeter *et al.*, 2015). These strategies included communication and shared decision-making training using a variety of approaches, from blended learning approaches to interactive booklets and group education (Coxeter *et al.*, 2015). A summary of these interventions can be found in Table 6.1. A more recent study

Table 6.1. Table of interventions to facilitate shared decision-making to address antibiotic use for acute respiratory infections in primary care. (From Coxeter *et al.*, 2015)

Author/Year	Intervention	Recipient / Where	What (materials)	What (procedures)	When and how much
Briel, 2006	Brief training programme in patient-centred communication	GPs / Not specified	Evidence-based guidelines for diagnosis and treatment of ARIs (updated, locally adapted and reviewed by local experts) distributed as a booklet	GPs were trained in elements of active listening, to respond to emotional cues and to tailor information given to patients. Physicians were introduced to a model (Prochaska, 1992) to identify patients' attitude and readiness for behaviour change.	Attendance at 1 x 6-hour seminar and 1 x 2-hour telephone call to give personal feedback prior to the trial start
Butler, 2012	Multifaceted flexible blended learning approach for clinicians	GPs and nurse practitioners / At the general practice	Summaries of research evidence and guidelines, web-based modules using video-rich material presenting novel communication skills, and a web-based forum to share experiences and views (see www.stemmingthetide.org for online component)	Intervention consists of 7 components: experiential learning, updated summaries of research evidence and guidelines; web-based learning in novel communication skills; practising consulting skills in routine care; facilitator-led practice-based seminar on practice-level data on antibiotic prescribing and resistance; reflections on own clinical practice, and a web-based forum to share experiences and views.	7 components (5 online, 1 face-to-face and 1 facilitator-led practice-based seminar) A booster module (6 to 8 months after completion of initial training) reinforced these skills

Continued

Table 6.1. Continued.

Author/Year	Intervention	Recipient Where	What (materials)	What (procedures)	When and how much
Cals, 2009	Enhanced communication skills training	GPs General practice	Pre- and post-workshop transcripts of simulated patients	Brief context-learning based workshop in small groups (5–8 GPs), preceded and followed by practice-based consultations with simulated patients. GPs reflected on own transcripts of consultations with simulated patients, which were also peer-reviewed by colleagues.	1 x 2-hour moderator-led small groups workshop, preceded and followed by practice-based consultation with simulated patients
Francis, 2009	Interactive booklet for parents and clinician training in its use	GPs and patients General practice, patients' houses	8-page booklet (now at www.whenshouldIworry.com); online training in use of the booklet included videos to demonstrate use of the booklet within a consultation, as well as audio feeds, pictures and links to study materials	Booklet given to parents to use in the consultation and as a take-home resource (no further details provided). Online training in the use of the booklet was provided to GPs: describing the content and aims of the booklet, and encouraging use within the consultation to facilitate use of specific communication skills.	1 x 40-minute online training module

Légaré, 2012	Shared decision-making programme (DECISION+2)	Family physicians (including teachers and residents) Family practice teaching units	Online tutorial and workshop included videos, exercises and decision aids to help physicians communicate to their patients the probability of bacterial ARIs and benefits/harms of antibiotic use. Decision aids were available in the consultation rooms in all family practice teaching units.	Online self-tutorial comprising 5 modules, 2-hour online tutorial followed by facilitator-led, on-site interactive workshops aimed to help physicians review and integrate concepts acquired during online training.	1 x 2-hour online tutorial, followed by 1 x 2-hour on-site interactive workshop. Participants had 1 month to complete the programme
Légaré, 2011	Multiple-component, continuing professional development programme in shared decision-making (DECISION+)	Family medicine groups (physicians and nurses) Family medicine groups	Workshops included videos (simulated consultations of usual care and SDM) and exercises (facilitators and barriers to SDM). GPs trained in the use of 5 decision support tools using video examples and group exercises. A booklet summarizing workshop content was provided to participants. Postcard reminders were sent.	Interactive workshops and related material, reminders of expected behaviours and GP feedback on agreement between their decisional conflict and that of their patients.	3 x 3-hour interactive workshops and related material, in addition to reminders of expected behaviours and GP feedback on agreement between their decisional conflict and that of their patients. DECISION+ conducted over 4 to 6 months

Continued

Table 6.1. Continued.

Author/Year	Intervention	Recipient Where	What (materials)	What (procedures)	When and how much
Little, 2013	Internet-based training in enhanced communication skills	GPs General practice	Interactive booklet for use by GPs within consultations Training supported by video demonstrations of consultation techniques	Online modules and an interactive booklet for use within consultations (group practices also appointed a lead GP to organize a structured meeting on prescribing issues).	Internet modules completed alone or in a group
Welschen, 2004	Group education meeting with consensus procedure and communication skills training	GPs/ pharmacists and their assistants, and patients Not described	Group consensus guidelines and patient waiting-room materials (poster/ leaflets)	Group education meeting with consensus procedure, with a summary, and guidelines mailed 1 month later to reinforce consensus; feedback on prescribing behaviour (post- and pre-intervention insurance claims data) and practice-level reporting of prescribing behaviours aligned with consensus reached; group education session for GP and pharmacists' assistants (Dutch guidelines and skills training in patient education); waiting-room educational material for patients.	1 x group education meeting with consensus procedure; 1 x 2-hour group education session for GP and pharmacists' assistants; monitoring and feedback of prescribing behaviour at 6 months post-intervention

by van Esch *et al.* (2018) in Holland also found that in primary care practices where SDM took place more frequently, GPs prescribed fewer antibiotics for adults under the age of 40 in preference-sensitive situations (where, according to guidelines, antibiotics could be considered but were not mandatory).

Other examples of stewardship in action have focused on the use of patient information. The TARGET Antibiotics Toolkit developed by the Public Health England Primary Care Unit (PHE, 2012), the Royal College of General Practitioners (RCGP, 2019) and the Antimicrobial Stewardship in Primary Care Group is a resource to help prescribers improve antibiotic prescribing in primary care (where the majority of antibiotics are prescribed). As part of this resource, a TARGET ('Treating Your Infection') leaflet was developed, which was designed to be shared with the patient and completed with them during the consultation (RCGP, 2019).

Case study 1

An older patient and his carer (his wife) have just been informed that he has *Clostridioides difficile.* He is still experiencing severe diarrhoea and is being cared for in a single room with en suite facilities within a medical ward. He was admitted 24 hours earlier with a suspected chest infection, which was treated unsuccessfully with three different antibiotics while at home. The patient is a little confused, so he does not appear to understand this diagnosis. However, his wife is very upset and angry. She believes that he has acquired *C. difficile* since being admitted to hospital and does not understand why his antibiotics must be changed again. She is also very frightened because she has read all about outbreaks in the newspaper. You need to go and speak with this patient, his wife and his grown-up son and daughter, who are also very upset by this diagnosis.

1. What are the potential barriers to effective communication in this situation?
2. What would you say to start this conversation?
3. What information would you give to this family?
4. What strategies would you take to engage with the family and develop a rapport?

Answers:

1. The potential barriers include:
   a. The family are frightened.
   b. The patient is confused and therefore not able to process information given to him.
   c. The patient's wife is upset and angry. She also does not understand the change in treatment. The patient's son and daughter are also angry.
   d. The patient's wife already has perceptions of the situation because of what she has read in the media. These perceptions may not reflect her husband's situation.

Continued

Case study 1 Continued.

2. How to start the conversation:
   a. Introduce yourself – your name and who you are.
   b. Reassure them that you are there to help them and to answer any questions they have.
   c. Ask them what they know and how they feel.
3. What information would you give?
   a. Give them the information that is important to *them* (regardless of how important you feel it is); you must ensure that they know their concerns are being taken seriously.
   b. Tell the family exactly what the infection is (don't use terms like 'a little bug'), and explain as clearly and simply as you can the key factors – causative factors, transmission, treatment, prevention, etc.
   c. Explain exactly what the treatment is, how long it will be needed, any side effects and what will happen when it stops.
   d. Explain how you will know if the infection has been successfully treated and any implications longer term.
   e. Give all the family the opportunity to ask questions and reiterate what you have said so that they have understood.
   f. Give them a contact should they have further questions at a later date.
4. Strategies:
   a. Ensure the environment is right.
   b. Ensure you have time to spend speaking with them for as long as they need.
   c. As above, introduce yourself.
   d. Listen more than you talk.
   e. Keep eye contact.
   f. Always tell the truth. Never lie. If you don't know, say you don't know but you will find out.

Case study 2

Mrs Campbell visits her GP with her five-year old daughter, Eve, who has a sore throat. The history suggests that Eve has had symptoms of a sore throat for 5 days. She has a cough but no fever and on examination her tonsils are not severely inflamed. The GP says that it is likely to be a viral infection and that she should continue to give Eve analgesics for the pain and that an antibiotic is not needed. Mrs Campbell is not happy with this management and explains that Eve received an antibiotic last year when she had a throat infection and they cleared it up immediately. She also adds that she cannot afford to be off work any longer and needs Eve to return to school.

1. How could the GP communicate better with Mrs Campbell?
2. What other strategies could be used?

Continued

Case study 2 Continued.

3. Are there any moral or psychological arguments for prescribing an antibiotic in this situation?
4. Can you think of any other pressures or influences that may make them consider prescribing an antibiotic? How could these be managed?
5. If Eve also had another symptom on the FeverPAIN or Centor score, e.g. fever (NICE, 2018), what could the GP do in this situation if they were still concerned that it was a viral infection?

Answers:

1. The GP could use good communication skills and shared decision-making as discussed in this chapter – sharing information, involving them in shared decision-making and listening to their concerns. Clear explanations should be provided about the difference between viral and bacterial infections, the use of national prescribing guidelines, the side effects of antibiotics (for the individual and for the public health) and the difference between Eve's recovery if she was or wasn't given antibiotics. Self-management advice should also be provided regarding over-the-counter medication, fluids and rest. Educational aids such as leaflets and waiting-room or surgery posters could also be used.
2. Safety-netting strategies could be employed such as organizing a return appointment for Mrs Campbell if Eve's symptoms don't improve, or delayed prescribing, along with advice about the likely length of symptoms.
3. There may be an argument that prescribing an antibiotic would decrease patient anxiety and, if bacterial, reduce symptoms by 1–2 days; however, the public health concern of AMR should be the priority. Advice about the possible side effects of antibiotics and their ineffectiveness against a viral infection should be stressed, along with advice about regular analgesia.
4. The fact that Eve received an antibiotic previously for a similar presentation (precedence for antibiotic prescribing) and the length of time of the consultation, i.e. how long the GP has to discuss everything with Mrs Campbell, could influence prescribing decisions.
5. The GP could again use safety-netting, i.e. use delayed prescribing and inform Mrs Campbell that if Eve's symptoms do not improve or get worse then she could start taking the antibiotics. These can be post-dated and can be given to Mrs Campbell or left at the local pharmacy to be collected at a later date.

Key Points

1. Person-centred care, where patients are actively involved in their own care, has many benefits, which include improved adherence to treatment and a better patient experience.
2. Evidence suggests that healthcare professionals do not always involve the patient and do not always provide the patient with the appropriate information about their infection and its treatment.

3. Healthcare professionals should actively listen to patients and provide them with appropriate information about their condition and treatment/advice.

4. Patient or carer expectations for an antibiotic is an influence on prescribers.

5. Prescribers should be prepared for this expectation but it should not influence their decision; rather it should influence the management strategies they adopt to manage the situation.

References

Ardalan, F., Bagheri-Saweh, M.I., Etemadi-Sanandaji, M., Nouris B. and Valiee, S. (2018) Barriers of nurse-patient communication from the nurses' point of view in educational hospitals affiliated to Kurdistan University of Medical Science. *Nursing Practice Today* 5(2), 290–298.

Almutairi, K.M. (2015) Culture and language differences as a barrier to provision of quality care by the health workforce in Saudi Arabia. *Saudi Medical Journal* 36(4), 425–431. DOI: 10.15537/smj.2015.4.10133

Beagley, L. (2011) Educating patients: understanding barriers, learning styles, and teaching techniques. *Journal of PeriAnesthesia Nursing,* 26(5), 331–337.

Berman, A.C. and Chutka, D.S. (2016) Assessing effective physician–patient communication skills: Are you listening to me, doc? *Korean Journal of Medical Education* 28(2), 243–249. DOI: 10.3946/kjme.2016.21

Boyd, S.E., Moore, L.S.P., Gilchrist, M., Costelloe, C., Castro-Sánchez, E., Franklin, B.D. and Holmes, A.H. (2017) Obtaining antibiotics online from within the UK: a cross-sectional study. *Journal of Antimicrobial Chemotherapy* 72(5). DOI: 10.1093/jac/dkx003

Brooks, L., Shaw, A., Sharp, D. and Hay, A.D. (2008) Towards a better understanding of patients' perspectives of antibiotic resistance and MRSA: a qualitative study. *Family Practice* 5, 341–348. Available at: https://www.ncbi.nlm.nih.gov/pubmed/18647956 (accessed 24 September 2019).

Butler, C.C., Kinnersley, P., Prout, H., Rollnick, S., Edwards, A., *et al*. (2001) Antibiotics and shared decision-making in primary care. *Journal of Antimicrobial Chemotherapy* 3, 435–440. Available at: https://academic.oup.com/jac/article/48/3/435/736084 (accessed 24 September 2019).

Carter, R.R., Sun, J. and Jump, R.L. (2016) A survey and analysis of the American public's perceptions and knowledge about antibiotic resistance. *Open Forum Infectious Diseases* 3. Available at: https://www.ncbi.nlm.nih.gov/pubmed/27382598 (accessed 24 September 2019).

Charani, E., Tarrant, C., Moorthy, K., Sevdalis, N., Brennan, L., *et al*. (2017) Understanding antibiotic decision making in surgery – a qualitative analysis. *Clinical Microbiology and Infection* 23, 752–760. Available at: https://www.sciencedirect.com/science/article/pii/S1198743X17301829 (accessed 24 September 2019).

Coco, A. and Mainous, A.G. (2005) Relation of time spent in an encounter with the use of antibiotics in pediatric office visits for viral respiratory infections. *Archives of Pediatrics and Adolescent Medicine* 159, 1145–1149. Available at: http://dx.doi.org/10.1001/archpedi.159.12.1145 (accessed 24 September 2019).

Coenen, S., Francis, N., Kelly, M., Hood, K., Nuttall, J., *et al*. (2013) Are patient views about antibiotics related to clinician perceptions, management and outcome? A multi-country study in outpatients with acute cough. *PLOS One* 8, e76691. DOI: 10.1371/journal.pone.0076691

Courtenay, M., Rowbotham, S., Lim, R., Deslandes, R., Hodson, K., *et al.* (2017) Antibiotics for acute respiratory tract infections: a mixed-methods study of patient experiences of non-medical prescriber management. *BMJ Open* 7, e013515. Available at: https://bmjopen.bmj.com/content/7/3/e013515 (accessed 24 September 2019).

Coxeter, P., Del Mar, C.B., McGregor, L., Beller, E.M. and Hoffmann, T.C. (2015) Interventions to facilitate shared decision making to address antibiotic use for acute respiratory infections in primary care. *Cochrane Database of Systematic Reviews* (11), CD010907. Available at: https://www.cochranelibrary.com/cdsr/doi/10.1002/14651858.CD010907.pub2/full (accessed 20 December 2019).

Currie, K., Melone, L., Stewart, S., King, C., Holopainen, A., *et al.* (2018) Understanding the patient experience of health care-associated infection: a qualitative systematic review. *American Journal of Infection Control* 8, 936–942. Available at: https://www.ajicjournal.org/article/S0196-6553(17)31292-0/fulltext (accessed 24 September 2019).

Davis, M.E., Tsai-Ling, L., Yhenneko, J.T., Davidson, L., Schmid, M., *et al.* (2017) Exploring patient awareness and perceptions of the appropriate use of antibiotics: a mixed-methods study. *Antibiotics (Basel)* 6(4), 23. DOI: 10.3390/antibiotics6040023

Dyche, L. and Swiderski, D. (2005) The effect of physician solicitation approaches on ability to identify patient concerns. *Journal of General Internal Medicine* 20(3), 267–270. Available at: https://www.ncbi.nlm.nih.gov/pubmed/15836531 (accessed 24 September 2019).

Ebdel, M.H., Lundgren, J. and Youngpairoj, S. (2013) How long does a cough last? Comparing patients' expectations with data from a systematic review of the literature. *The Annals of Family Medicine* 11(1), 5–13. DOI: 10.1370/afm.1430

Father, C.P. and Stevens, S. (2008) Improving the patient's experience. *Community Eye Health* 21(68) 55–57. Available at: https://www.cehjournal.org/article/improving-the-patients-experience/ (accessed 12 March 2019).

Fletcher-Lartey, S., Yee, M., Gaarslev, C. and Khan, R. (2016) Why do general practitioners prescribe antibiotics for upper respiratory tract infections to meet patient expectations: a mixed method study. *BMJ Open* 6, e012244. DOI: 10.1136/bmjopen-2016-012244

Francis, N.A., Gillespie, D., Nuttall, J., Hood, K., Little, P., *et al.* (2012) Delayed antibiotic prescribing and associated antibiotic consumption in adults with acute cough. *British Journal of General Practice* 62(602), e639–e646.

Gaarslev, C., Yee, M., Chan, G., Fletcher-Lartey, S. and Khan, R. (2016) A mixed methods study to understand patient expectations for antibiotics for an upper respiratory tract infection. *Antimicrobial Resistance and Infection Control* 5, art. 39. DOI: 10.1186/s13756-016-0134-3

Ham, C., Charles, A. and Wellings, D. (2018) Shared responsibility for health: the cultural change we need. Available at: https://www.kingsfund.org.uk/publications/shared-responsibility-health (accessed 27 February 2019).

Harbarth, S., Balkhy, H.H., Goossens, H., Jarlier, V., Kluytmans, J., *et al.* (2015) Antimicrobial resistance: one world, one fight! *Antimicrobial Resistance and Infection Control* 4, 49. Available at: https://www.ncbi.nlm.nih.gov/pmc/articles/PMC4652432/ (accessed 24 September 2019).

Hoffman, T.C., Légaré, F., Simmons, M.B., McNamara, K., McCaffery, K., *et al.* (2014) Shared decision making: what do clinicians need to know and why should they bother? *The Medical Journal of Australia* 201(1), 35–39. Available at: https://www.mja.com.au/journal/2014/201/1/shared-decision-making-what-do-clinicians-need-know-and-why-should-they-bother (accessed 24 September 2019).

Houts, P.S., Doak, C.C., Doak, L. and Loscalzo, M.J. (2006) The role of pictures in improving health communication: a review of research on attention, comprehension, recall and adherence. *Patient Education and Counseling* 61(2), 173–190. DOI: 10.1016/j.pec.2005.05.004

Jifar, A. and Ayele, Y. (2018) Assessment of knowledge, attitude and practice toward antibiotic use among Harar City and its surrounding community, Easter Ethiopia. *Interdisciplinary Perspectives on Infectious Diseases*, art. 8492740. Available at: https://doi.org/10.1155/2018/8492740 (accessed 24 September 2019).

Jose, J., Jimmy, B., AlSabahi, A.G.M.S. and Sabei, G.A.A. (2013) A study assessing public knowledge, belief and behavior of antibiotic use in an Omani population. *Oman Medical Journal* 28(5), 324–330. Available at: https://www.ncbi.nlm.nih.gov/pmc/articles/PMC3769127/ (accessed 24 September 2019).

Khan, D. (2010) Fixing the communications failure. *Nature* 463, 296–297. Available at: https://ssrn.com/abstract=1630002 (accessed 12 March 2019).

King, L.M., Fleming-Dutra, K.E. and Hicks, L.A. (2018) Advances in optimizing the prescription of antibiotics in outpatient settings. *BMJ* 363, k3047. DOI: https://doi.org/10.1136/bmj.k3047

Lee, C., Lee, J.H., Kang, L., Jeong, B.C. and Lee, S.H. (2015) Educational effectiveness, target and content for prudent antibiotic use. *BioMed Research International*, 214021. Available at: https://www.ncbi.nlm.nih.gov/pmc/articles/PMC4402196/ (accessed 12 March 2019).

Linder, J.A., Singer, D.E. and Stafford, R.S. (2003) Association between antibiotic prescribing and visit duration in adults with upper respiratory tract infections. *Clinical Therapeutics* 25, 2419–2430. Available at: https://www.ncbi.nlm.nih.gov/pubmed/14604741 (accessed 24 September 2019).

Lum, E.P.M., Page, K., Whitty, J.A., Doust, J. and Graves, N. (2018) Antibiotic prescribing in primary healthcare: dominant factors and trade-offs in decision-making. *Infection, Disease and Health* 23, 74–86. Available at: https://doi.org/10.1016/j.idh.2017.12.002 (accessed 24 September 2019).

Marcus, C. (2014) Strategies for improving the quality of verbal patient and family education: a review of the literature and creation of the EDUCATE model. *Health Psychology and Behavioural Medicine* 2, 482–495. DOI: 10.1080/21642850.2014.900450

Mauksch, L. (2017) Questioning a taboo: physicians' interruptions during interactions with patients. *JAMA* 317, 1021–1022. Available at: https://www.ncbi.nlm.nih.gov/pubmed/28291896 (accessed 24 September 2019).

McCormack, B. and McCance, T. (2016) Person-Centred Practice in Nursing and Health Care: Theory and Practice, 2nd Edition. Wiley, Chichester, UK.

McKay, R., Mah, A., Law, M.R., McGrail, K. and Patrick, D.M. (2016) Systematic review of factors associated with antibiotic prescribing for respiratory tract infections. *Antimicrobial Agents and Chemotherapy* 60, 4106–4118. DOI: 10.1128/AAC.00209-16

Mustafa, M., Wood, F., Butler, C.C. and Elwyn, G. (2014) Managing expectations of antibiotics for upper respiratory tract infections: a qualitative study. *Annals of Family Medicine* 12(1), 29–36.

National Institute for Health and Care Excellence (NICE) (2014) Pneumonia in adults: diagnosis and management. Available at: https://www.nice.org.uk/guidance/cg191 (accessed 4 January 2019).

NICE (2015) Antimicrobial stewardship: systems and processes for effective antimicrobial medicine use. Available at: https://www.nice.org.uk/guidance/ng15 (accessed 4 January 2019).

NICE (2018) Sore throat (acute): antimicrobial prescribing. NICE guideline 84. Available at: https://www.nice.org.uk/guidance/ng84 (accessed 4 January 2019).

Ospina, N.S., Phillips, K.A., Rodriguez-Gutierrex, R., Castaneda-Guarderas, A., Gionfriddo, M.R., *et al.* (2019) Eliciting the patient's agenda – secondary analysis of recorded clinical encounters. *Journal of General Internal Medicine* 34, 36–40. DOI: 10.1007/s11606-018-4540-5

Panagioti, M., Skevington, S., Hann, M., Howells, K., Blakemore, A., *et al.* (2018) Effect on health literacy on the quality of life of older patients with long-term conditions: a large cohort study in UK general practice. *Quality of Life Research* 27, 1257–1268. DOI: 10.1007/s11136-017-1775-2

Pechère, J.C. (2001) Patients' interviews and misuse of antibiotics. *Clinical Infectious Diseases* 15 (Suppl 3), S170–S173. DOI: 10.1086/321844

Public Health England (2012) Training Resources: *TARGET Antibiotics Toolkit*. Available at: www.rcgp.org.uk/targetantibiotics (accessed 4 January 2019).

Rawson, T.M., Moore, L.S.P., Hernandez, B., Castro-Sánchez, E., Charani, E., *et al.* (2016) Patient engagement with infection management in secondary care: a qualitative investigation of current experiences. *BMJ Open* 6, e011040. DOI: 10.1136/bmjopen-2016-011040

Rawson, T.M., Moore, L.S.P., Castro-Sánchez, E., Charani, E., Hernandez, B., *et al.* (2018) Development of a patient-centred intervention to improve knowledge and understanding of antibiotic therapy in secondary care. *Antimicrobial Resistance and Infection Control* 7, art. 43. Available at: https://doi.org/10.1186/s13756-018-0333-1 (accessed 24 September 2019).

Rowbotham, S., Chisholm, A., Moschogianis, S., Chew-Graham, C., Cordingly, L., *et al.* (2012) Challenges to nurse prescribers of a no-antimicrobial prescribing strategy for managing self-limiting respiratory tract infections. *Journal of Advanced Nursing* 68(12), 2622–2632. DOI: 10.1111/j.1365-2648.2012.05960.x

RCGP (Royal College of General Practitioners) (2019) *TARGET Antibiotic Toolkit*. Available from: https://www.rcgp.org.uk/clinical-and-research/resources/toolkits/target-antibiotic-toolkit.aspx (accessed 12 March 2019).

Scholl, I., Zill, J.M., Härter, M. and Dirmaier, J. (2014) An integrative model of patient-centeredness – a systematic review and concept analysis. *PLOS One* 9, e107828. Available at: https://doi.org/10.1371/journal.pone.0107828 (accessed 24 September 2019).

Sharp, A.L., Shen., E., Kanter, M.H., Berman, L.J. and Gould M.K. (2017) Low-value antibiotic prescribing and clinical factors influencing patient satisfaction. *American Journal of Managed Care* 23(10), 589–594.

Sørensen, K., Van den Broucke, S., Pelikan, J.M., Fullam, J., Doyle, G., *et al.* (2013) Measuring health literacy in populations: illuminating the design and development process of the European Health Literacy Survey Questionnaire (HLS-EU-Q). *BMC Public Health* 13, art. 948.

Stans, S.E.A., Dalemans, R.J.P., de Witte, L.P., Smeets, H.W.H. and Beurskens, A.J. (2016) The role of the physical environment in conversations between people who are communication vulnerable and health-care professionals: a scoping review. *Disability and Rehabilitation* 39, 2594–2605. DOI: 10.1080/09638288.2016.1239769

Tarrant, C., Colman, A.M., Jenkins, D.R., Mehtar, S., Perera, N. and Krockow, E.M. (2019) Optimizing antibiotic prescribing: collective approaches to managing a common-pool resource. *Clinical Microbiology and Infection* 72(5), 1521–1528.

Tong, S., Pan, J., Lu, S. and Tang, J. (2018) Patient compliance with antimicrobial drugs: a Chinese survey. *American Journal of Infection Control* 46, e25–e29.

Tonkin-Crine, S., Anthierens, S., Francis, N.A., Brugman, C., Fernandez-Vandellos, P., *et al.* (2014) Exploring patients' views of primary care consultations with contrasting interventions for acute cough: a six-country European qualitative study. *NPJ Primary Care Respiratory Medicine* 24, 14026. DOI: 10.1038/npjpcrm.2014.26

Vallin, M., Polyzoi, M., Marrone, G., Rosales-Klintz, S., Wisell, K.T., *et al.* (2016) Knowledge and attitudes towards antibiotic use and resistance – a latent class analysis of a Swedish population-based sample. *PLOS ONE* 11(4), e0152160. Available at: https://doi.org/10.1371/journal.pone.0152160 (accessed 24 September 2019).

Van Esch, T.E.M., Brabers, A.E.M., Hek, K., van Dijik, L., Verheij, R.A., *et al.* (2018) Does shared decision-making reduce antibiotic prescribing in primary care? *Journal of Antimicrobial Chemotherapy* 73(11), 3199–3205. DOI: 10.1093/jac/dky321

Viney, K.A., Johnson, P., Tagaro, M., Fanai, S., Linh, N., *et al.* (2014) Tuberculosis patients' knowledge and beliefs about tuberculosis: a mixed methods study from the Pacific Island national of Vanuatu. *MBC Public Health* 14, 467. Available at: https://www.ncbi.nlm.nih.gov/pubmed/24885057 (accessed 24 September 2019).

Welschen, I., Kuyvenhoven, M., Hoes, A. and Verheij, T. (2004) Antibiotics for acute respiratory tract symptoms: patients' expectations, GPs' management and patient satisfaction. *Family Practice* 21(3), 234–237. Available at: https://doi.org/10.1093/fampra/cmh303 (accessed 24 September 2019).

Interprofessional Collaborative Practice

7

Fiona Gotterson[1],* and Elizabeth Manias[2]

[1]*PhD Fellow, National Centre for Antimicrobial Stewardship (NCAS), University of Melbourne, Victoria, Australia;* [2]*Deakin University, School of Nursing and Midwifery, Centre for Quality and Patient Safety Research, Institute for Health Transformation, Burwood, Australia; The University of Melbourne, Department of Medicine, Royal Melbourne Hospital, Parkville, Australia*

Objective: For the student to understand how different professions collaborate in relation to how they contribute to antimicrobial stewardship and to describe the nurse role in interprofessional collaboration to promote appropriate antimicrobial use.

Introduction

Nurses represent the biggest percentage of the health workforce and practise in a range of settings across the continuum of care, and thus are responsible for actions that contribute to the safety and quality of patient care (World Health Organization (WHO), 2016). Moreover, nurses have a focus on patient-centred healthcare, which is associated with better patient outcomes (Hughes, 2008; Ewers *et al.*, 2017). In the case of antimicrobial stewardship (AMS), there have been calls for wider consideration of the nurse's role, and for nurses to be recognized as legitimate contributors to AMS team efforts (Edwards *et al.*, 2011; Olans *et al.*, 2016; American Nursing Association and Centers for Disease Control and Prevention, 2017; Castro-Sánchez *et al.*, 2017). Interprofessional collaboration and teamwork are widely acknowledged as essential to safe quality patient care, and to AMS, and have been recommended as a domain of an undergraduate competency framework for AMS (Courtenay *et al.*, 2018), but novice health professionals may feel inadequately prepared or lack confidence to practise collaboratively with professionals from other disciplines (Foronda *et al.*, 2016; Levett-Jones *et al.*, 2018). This chapter explores influences on interprofessional collaboration, relevance of interprofessional collaboration for AMS, potential strategies for establishing good

*Corresponding author: fgotterson@student.unimelb.edu.au

interprofessional working relationships, and tools and resources that may support enhanced collaboration between nurses and other health professionals.

Why Interprofessional Collaboration?

What is interprofessional collaborative practice?

According to the WHO, collaborative practice happens when 'multiple health workers from different professional backgrounds work together with patients, families, carers and communities to deliver the highest quality of care' (WHO, 2010). The terms 'interprofessional' (or interdisciplinary), 'multidisciplinary' and 'teamwork' are interrelated, but each has different meanings (D'Amour *et al.*, 2005; Hall, 2005; Reeves *et al.*, 2018). A multidisciplinary effort, for example, refers to circumstances where individuals work on the same project, but in parallel with one another, each with separate goals and responsible for activities relevant to their discipline (D'Amour *et al.*, 2005). Teamwork can enhance collaboration, but having health professionals work in teams will not necessarily result in collaboration (D'Amour *et al.*, 2005; Reeves, 2012).

An interprofessional approach, by comparison, involves professionals from different disciplines working together towards a shared goal, so that patient care interventions are integrated (D'Amour *et al.*, 2005; Petri, 2010). For the purposes of this chapter, we use the following definition, adapted from Petri (2010, page 79): 'Interprofessional collaboration is an interpersonal process characterised by health professionals from multiple disciplines, with shared objectives, decision-making, responsibility, and power, working together to solve patient care problems.' Collaboration requires persons to set aside individual intentions, and rather than argue over role boundaries and who should do what, to establish a common understanding about the goals and objectives for patient care and about how different health professionals will work together to achieve the identified goals (Petri, 2010). A shared view about what patient-centredness means, and about the patient's role as a member of the interprofessional team, is also necessary (D'Amour *et al.*, 2005). Central attributes of interprofessional collaboration are 'mutual trust and respect, effective and open communication, and awareness and acceptance of the roles, skills, and responsibilities of the participating disciplines' (Petri, 2010, page 79).

Established benefits of interprofessional collaboration include enhanced patient satisfaction and less fragmented care (D'Amour *et al.*, 2005; Petri, 2010; Reeves *et al.*, 2017). The WHO Framework for Action on Interprofessional Education and Collaborative Practice identifies several studies that demonstrate the potential for collaborative practice to reduce length of stay, and clinical error, and improve health outcomes (WHO, 2010). There is consistent evidence which shows that breakdowns in effective communication and collaboration are significant contributing factors to serious adverse events in healthcare (Levett-Jones *et al.*, 2018). For organizations and healthcare systems, collaboration is also associated with decreased costs and more efficient use of health resources (Petri, 2010; Reeves *et al.*, 2017). There are benefits, too, for health professionals; enhanced job satisfaction is frequently associated with interprofessional collaboration (Petri, 2010). Job satisfaction impacts work engagement, which impacts patient care; when health professionals are not engaged in their work, and have reduced job satisfaction, their capacity to deliver good care is compromised (Stein-Parbury, 2018).

Interprofessional collaboration and AMS

Antimicrobials are essential drugs for the treatment and prevention of serious infections. But overuse or misuse of antimicrobials may also be associated with adverse consequences for individuals, and for populations. Given the potential for antimicrobials to benefit, but paradoxically to also cause harm, AMS is primarily a patient safety activity, comprising several complementary activities that occur across the patient care pathway, from diagnosis, to the act of prescribing, to administration and patient monitoring, to discharge. Although AMS strategies must be tailored to reflect local priorities, and resources, stewardship activities are relevant in all healthcare settings where antimicrobials are used (Cosgrove *et al.*, 2014; Barlam *et al.*, 2016) including acute care hospitals, aged and long-term care, outpatient clinics and in primary care. Different health professionals from several disciplines are involved in AMS, and each brings different knowledge, skills and perspectives (ACSQHC, 2018). Like any patient safety endeavour, AMS requires health professionals to practise collaboratively together if best patient and health outcomes are to be achieved.

For AMS, collaboration requires team members to understand that AMS is not an intervention that is owned or directed by a single individual, but that responsibility for safe use of antimicrobials is shared amongst all members of the health team, which includes the patient. All members of the interprofessional team need a common understanding of the goals of antimicrobial management, the expected outcomes of treatment, and an understanding of how each team member contributes to good antimicrobial management within the context of their role. Patients, as members of the team, should also understand individual team member contributions (D'Amour *et al.*, 2005). Knowing and understanding results of local surveillance activities is also important. Surveillance of antimicrobial usage and resistance, followed by feedback on results of this surveillance to health professionals, is a core AMS activity (Barlam *et al.*, 2016). Information obtained from surveillance activities guides practice and clinical decision-making, and underpins local AMS policy development. It is widely acknowledged that if people receive feedback about their performance and practice it will encourage them to make changes, thus driving improvement (Charani *et al.*, 2011; Davey, 2015). Providing feedback about current antimicrobial usage and resistance patterns, to show where improvement may be needed, to everyone involved in AMS can also help give the team a shared focus, thereby encouraging collaboration.

AMS team member roles
Early AMS guidelines described a model led by infectious diseases specialists, pharmacists and microbiologists (Dellit *et al.*, 2007). Nurses were ostensibly excluded, despite their established roles in medication management,

infection prevention and patient safety (Olans *et al.*, 2016). Learning from implementation of AMS programmes in settings outside of tertiary referral hospitals, and a growing appreciation that a model based around immediate availability of experts may not be possible in all settings, has led to an understanding that more diversity in AMS teams was needed to enable programmes to meet local patient need (Cosgrove *et al.*, 2014; Castro-Sánchez *et al.*, 2017).

Leadership by an infectious diseases physician and pharmacist remains the gold standard for AMS (Barlam *et al.*, 2016); however, organizations and experts now recommend a wider collaborative approach, which includes nurses (American Nursing Association and Centers for Disease Control and Prevention, 2017; Castro-Sánchez *et al.*, 2017; Australian Commission on Safety and Quality in Health Care (ACSQHC), 2018). Roles for nurses in AMS have been proposed (see Chapters 4 and 8). Correspondingly, there has been increased understanding about the importance of engaging with patients and involving them in decision-making to optimize AMS efforts and to encourage more patient-centred AMS (Ewers, 2017). Nurses can have a significant role in empowering and enabling patients in understanding antimicrobial use and involvement in decisions about their care, which is in line with nursing's focus on patient-centred care and advocacy.

There are positive findings from research exploring the benefits of nurse involvement in AMS activities, which have relevance for interprofessional collaboration. For example:

- In Australian acute care wards, results of a quasi-experimental study whereby senior nurses collaborated with medical and pharmacy staff to deliver nursing education, focusing on promotion of intravenous (IV) to oral antibiotic switch in acute care settings, demonstrated an overall improvement in nurses' AMS knowledge following the intervention, and a reduction in the number of IV-line days for patients in three of the six wards that were included in the study (Gillespie *et al.*, 2013).
- In South Africa, collaboration between pharmacists and nurses led to improved timing of antimicrobial administration for urgent antimicrobial treatment (Messina *et al.*, 2015).
- In aged care settings, two studies undertaken in the USA (Zabarsky *et al.*, 2008) and Australia (Stuart *et al.*, 2015) reported significant reductions in total days of antimicrobials prescribed following interventions that involved nursing staff.

The value of nurses collaborating with health professionals from other disciplines on AMS has also been highlighted in qualitative research. Health professionals from different disciplines have identified that nurses often facilitate communication between patient and doctor, or between doctor and pharmacists, which can help build rapport and promote continuity of care (Cotta *et al.*, 2015). In ICU settings, two separate studies have shown how nurses are perceived by doctors, pharmacists and microbiologists to be effective in working collaboratively, and in communicating *within* the nursing team, but also with health professionals from other disciplines of medicine, microbiology and pharmacy, and to contribute meaningfully about patient progress, treatment plans and antimicrobial management (Rout and Brysiewicz, 2017; Jeffs *et al.*, 2018).

In the UK, a consensus set of competency statements has been developed to support health professionals to understand and engage with AMS in their clinical practice roles (Courtenay *et al.*, 2018). Descriptors within Domain 6 of these competency statements provide direction on the expectation for collaborative practice in AMS (see Box 7.1).

Castro-Sánchez *et al.* (2017) have proposed a model for AMS collaboration, which provides a useful reference point to show where different professional roles connect, and where opportunities for interprofessional collaboration lie (Fig. 7.1).

Box 7.1. Domain 6 competencies. (Courtenay *et al.*, 2018)

---

1. Demonstrate an understanding of the roles, responsibilities and competencies of other health professionals involved in antimicrobial treatment policy decisions.
2. Explain why it is important that healthcare professionals, involved in the delivery of antimicrobial therapy (including prescription, delivery and supply), have a common understanding of antimicrobial treatment policy decisions, the quantity of antimicrobial use and effective patient/client outcomes.
3. Establish collaborative communication principles and actively listen to other professionals and patients/carers involved in the delivery of antimicrobial therapy.
4. Communicate effectively to ensure common understanding of care decisions.
5. Develop trusting relationships with patients/carers and other health-/social care professionals.
6. Use information and communication technology effectively to improve interprofessional patient-centred care.

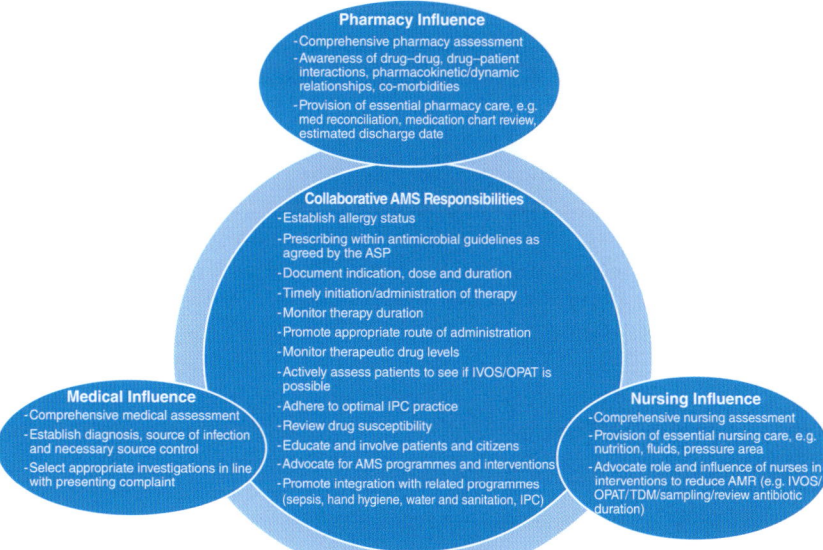

Fig. 7.1. Potential for collaboration between team members as part of an AMS programme. ASP, antimicrobial stewardship programme; IVOS, intravenous-oral switch; OPAT, Outpatient Antibiotic Therapy; TDM, Therapeutic Drug Monitoring.

Barriers to and Facilitators of Successful Interprofessional Collaboration

Barriers

Despite the obvious benefits, evidence shows that collaboration between different health professional groups does not always happen in practice (Tang *et al.*, 2013; Wilson *et al.*, 2016; Reeves *et al.*, 2017; Levett-Jones *et al.*, 2018). Nurses' capacity to be involved in AMS and to collaborate with others around antimicrobial management, can be either facilitated or prohibited by professional and organizational cultures, nurses' knowledge and confidence in antimicrobial use, their perceptions of their role and how others see the nurse role. Awareness of potential barriers to interprofessional collaboration can help in understanding why certain behaviours may occur, why collaborative practice may be difficult, and can encourage self-awareness and reflection about individual attitudes, behaviours toward other professions and working collaboratively.

Professional values, hierarchies

Hall (2005, page 188) explains that 'each health care profession has a different culture, incorporating values, beliefs, attitudes, customs and behaviours', and this influences health professionals' views of their role, the role of others and of the patients or consumers they provide care to (McCallin, 2001; D'Amour *et al.*, 2005; Hall, 2005).

Similarly, values and approaches to healthcare taught during undergraduate education differ between disciplines. For example, nurses are educated to work collectively in nursing teams (Hall, 2005), view the patient from a holistic perspective and see themselves as patient advocates (Raphael-Grimm, 2014). Doctors, by comparison, may focus more on individual expertise, have a more objective perspective of patient care (Hall, 2005; Curtis *et al.*, 2011; Raphael-Grimm, 2014) and may perceive collaboration differently to nurses (Hall, 2005; Tang *et al.*, 2013).

Such differences between disciplines can influence interprofessional collaboration. For example, one observational study involving doctors, nurses and pharmacists in an Australian public teaching hospital involved semi-structured interviews with 21 participants, and observations of 56 individuals working in acute care settings. The study revealed different attitudes towards medication management, and in communication behaviours, between the professional groups. Observations showed that the involved health professionals tended to work *alongside* one another, instead of *with* one another, and that when interprofessional communication happened, it was mostly task-focused (Rixon *et al.*, 2015). Differences in interprofessional communication styles may also be evident in electronic medical records (Bardach *et al.*, 2017).

Differences between professions are further complicated by the fact that health professions have historically been dominated by gender and hierarchy, led by medical staff (Broom *et al.*, 2016). Medical dominance can contribute to nurses' disempowerment and lack of confidence, thereby limiting their engagement and capacity for collaboration (Hall, 2005; Tang *et al.*, 2013; Foronda *et al.*, 2016; Stein-Parbury, 2018). Medical hierarchy can be a stronger determinant of anti-microbial prescribing behaviour than evidence-based guidelines or policies. This 'prescribing etiquette' (Charani *et al.*, 2013), whereby junior doctors see following the direction and prescribing preferences of senior doctors as more important than guidelines or policies, may limit the extent to which health professionals, including nurses, will question senior prescriber decisions, thus limiting collab-oration. Findings of qualitative research exploring the potential for nurse input into specific AMS activities shows that nurses in some acute care settings perceive questioning prescriber decisions will be unacceptable to medical staff (Fisher *et al.*, 2018; Monsees *et al.*, 2018). Indeed, nurses may work around established hier-archies, and even attempt to circumvent AMS policies, rather than question pre-scriber decisions, to influence antimicrobial management and prescribing (Broom *et al.*, 2016). While such actions may be motivated by precepts of patient-centred care and advocacy, this clearly impacts collaboration. By avoiding open discussion about antimicrobial management, the opportunity to foster collaborative relation-ships with medical staff is lost.

Organizational and workplace cultures

Just as professional and cultural differences impact collaboration, so too can the organizational culture. Organizational culture refers to the 'overarching culture of an organisation, including consistent practices, beliefs and attitudes'; for example, within an organizational setting (Braithwaite *et al.*, 2017, page 2). Workplace cul-ture is more about the culture experienced more directly by patients and health professionals at the level of day-to-day practice; for example, within a hospital ward (Manley *et al.*, 2011). The culture of an organization or workplace influences health professional satisfaction and engagement, and the extent to which care is person-centred, clinically effective and continually improving (Manley *et al.*, 2011); thus culture influences collaboration. Positive organizational and workplace cultures have also been found to be associated with positive patient outcomes, including reduced hospital-acquired infections and mortality rates (Braithwaite *et al.*, 2017).

A recent study by Charani *et al.* (2019) highlights the implications of organ-izational and professional cultures for AMS (Box 7.2).

Knowledge, confidence, role perceptions

Nurses and other health professionals may feel inadequately prepared to confi-dently communicate and collaborate with health professionals from other dis-ciplines (Tang *et al.*, 2013; Wilson *et al.*, 2016; Levett-Jones *et al.*, 2018). Equally, if an individual nurse is unclear about their role, and the role of others, within the context of the health team, this can detract from their capacity to engage in collaboration, as time and energy may be spent attempting to clarify roles (Wilson *et al.*, 2016).

Box 7.2. Impact of culture on AMS.

Charani *et al.* (2019) investigated the development and implementation of AMS programmes in health settings across five countries, representing different cultures, health systems and facing different implementation challenges. They found collaboration in AMS is only practised in a limited number of countries, and, in some countries, an absence of collaboration is associated with prevalent and ingrained medical hierarchies. However, even within countries where medical domination was prevalent, some individual organizational cultures diverged from the prevailing national culture. Interprofessional collaboration was evident in those organizations that had been acknowledged as centres of excellence, and held a reputation for innovation and leading 'good' AMS programmes. The authors stated that '…with the right leadership and drive, cultural norms can be challenged and changed over time', and concluded that this was more likely to be achieved in organizations that exhibit a more collectivist approach to teamwork and decision-making (Charani *et al.*, 2019).

Similarly, confidence in one's own *clinical* knowledge and skills may also impact an individual's capacity to engage in meaningful collaboration. Concerning AMS, research findings suggest mixed understanding among nurses about recognizing infection, infection management and about antimicrobials and AMS strategies (Zabarsky *et al.*, 2008; Gillespie *et al.*, 2013; Olans *et al.*, 2015; Wilson *et al.*, 2017). Nurses' views about their AMS role also vary. Nurses expect to be involved in AMS, and see they are patient advocates, but there are mixed views about what their involvement looks like in practice. Although some are keen for more active involvement in specific point-of-care interventions, such as IV to oral switch, if supported through education (Carter *et al.*, 2018; Fisher *et al.*, 2018; Monsees *et al.*, 2018), others perceive AMS to be mainly about prescribing rather than an interprofessional activity, and have clear views about the limits of their involvement (Charani *et al.*, 2013; Broom *et al.*, 2016; Fisher *et al.*, 2018). But maintaining a position about role boundaries is not necessarily in keeping with the objective of collaboration, where the focus is on sharing responsibility, planning and decision-making to enable optimal patient management and outcomes. Nor is this in keeping with the goals of AMS, which is about good antimicrobial management and practice across the continuum of care.

Perceptions of others about the nurse role is another potential barrier to collaborative practice in AMS. For example, not all professional groups will understand nurses' professional motivations as patient advocates, or their communication style (Foronda *et al.*, 2016). These reservations may limit trust, which is based on whether others perceive that an individual holds the necessary competence to be of value to the team, and is a requisite for meaningful collaboration (McCallin, 2001). This emphasizes the significance of establishing, between disciplines, shared understanding about roles, values and goals for patient care.

Facilitators

To develop skills in collaboration, health professionals need to learn and work together in meaningful ways (McCallin, 2001). Interprofessional education, use of guidelines and other tools and resources at the point of care, interprofessional rounds and meetings, and organizational and clinical leadership may all facilitate improved collaboration.

Reflective activity

As you read, reflect on your experiences as a student or novice nurse. Which of the following strategies and tools have you seen in practice, or participated in? Make notes about your experience of these, and any learnings or benefits for your practice as a result.

Interprofessional education (IPE)

Interprofessional education (IPE) refers to '…an intervention where the members of more than one health or social care profession, or both, learn interactively together, for the explicit purpose of improving interprofessional collaboration or the health/wellbeing of patients/clients, or both.' (Reeves *et al.*, 2013, page 2). IPE may be delivered in different modes, such as through face-to-face workshops, simulation activities, or via e-technology or online learning (World Health Organization, 2010), and is associated with improved communication and teamworking skills, increased appreciation and understanding of professional roles and values, and enhanced critical reflection skills (Foronda *et al.*, 2016). It has also been shown to encourage the shared focus on patient-centred care (Levett-Jones *et al.*, 2018). Competency statements have been developed to support interprofessional education and to guide disciplines to implement collaborative practices (Thistlethwaite *et al.*, 2014). There have been positive outcomes of interprofessional education activities (World Health Organization, 2010; Reeves *et al.*, 2013; Levett-Jones *et al.*, 2018), which indicates that collaborative practice skills can be learned.

One example of an online learning resource for AMS that targets all disciplines, and allows for communication between participants through online discussion forums, is the Antimicrobial Stewardship: Managing Antibiotic Resistance online course at https://www.futurelearn.com/courses/antimicrobial-stewardship (accessed 20 November 2019). The course outlines different roles and responsibilities in AMS and discusses different communication strategies. Participants completing the course can discuss and reflect on their perspective roles via the online discussion forum.

Guidelines, protocols, checklist and clinical pathways

Evidence-based guidelines, clinical pathways, checklists, posters and other point-of-care materials may support interprofessional collaboration and improve clinical processes and outcomes. Prescribing guidelines, for instance, are an essential component of AMS programmes; describing evidence-based best practice for antimicrobial prescribing, and 'a standard for prescribing in situations that are not explicitly described' within available guidance (ACSQHC, 2018, p. 81). Visual aids such as posters, pocket cards, clinical care pathways and care bundles, and checklists that are developed for us at the point of care can support use of guidelines in actual practice. For example:

- Use of a nurse-driven sepsis protocol incorporating a care bundle and complemented by education and feedback on performance against key indicators led to improvement in the early recognition and time to antimicrobial treatment of patients in emergency departments with sepsis. Empowering nurses to act on assessment findings was seen to facilitate a more multidisciplinary approach to sepsis management, thus improving quality of care (Tromp *et al.*, 2010).
- A prospective multi-centre study instigated across 33 hospitals in South Africa improved time from prescription to administration ('hang time') of first dose antimicrobials for sepsis. Activities included pharmacist-led education to doctors, nurses and pharmacists, posters displayed at the point of care, reminder emails to prescribers and weekly feedback to all health professionals on progress in achieving improvements in hang time. Hang time for first-dose antimicrobials increased from 41% (164/398) compliant before the intervention to 78.4% compliant after the intervention ($p<0.0001$). The authors of the study considered that this project provided a catalyst for collaboration among doctors, nurses, pharmacists and management, which could facilitate other AMS initiatives to be established within the hospital network (Messina *et al.*, 2015).

Interprofessional rounds, meetings

Having diverse groups of health professionals involved in interprofessional patient care rounds or meetings can support interprofessional collaboration and may improve use of health resources (Tang *et al.*, 2013; Reeves *et al.*, 2017). Defined as: encounters including 2 physicians plus a nurse or other care provider discussing the case at the patient's bedside (Gonzalo *et al.*, 2014, page 647), interprofessional rounds can enhance interprofessional relationships, increasing nurses' confidence in communicating with doctors (Tang *et al.*, 2013), and provide an opportunity for health professionals to understand and clarify issues, and help develop a shared understanding of patient care (Sharma and Klocke, 2014; Robert Wood Johnson Foundation, 2015). AMS team rounds are an example of an interprofessional round, and routinely involve core members of the local AMS team. Including nurses in interprofessional rounds, sharing results from audit and feedback processes and involving nurses in the planning of AMS activities were factors believed to encourage nurse involvement in AMS in Canadian intensive care units (Jeffs *et al.*, 2018).

Interprofessional meetings are another forum to promote interprofessional collaboration (Reeves *et al.*, 2017). Specific to AMS, involvement in interprofessional meetings where feedback on surveillance results are given offers nurses an opportunity to improve their understanding of local data and about what is happening locally in relation to antimicrobial usage, and to identify where they might be able to contribute and add value to local AMS activities. With this understanding, nurses, very likely, will feel more confident to discuss progress on AMS with other healthcare professionals and with patients. Guidelines recommend nurses be included on committees and working groups for development of AMS policy (PHE, 2011; ACSQHC, 2018). Participating in interprofessional meetings when policies and guidelines are being reviewed or developed, or when quality-improvement activities are being planned, can also help facilitate shared interprofessional learning and collaboration.

Surveillance of antimicrobial usage is often undertaken by a small interdisciplinary team; for example, a doctor, infection control professional or consultant, a nurse, or a pharmacist, or any combination of these. Taking opportunities to become involved in the process of surveillance may help nurses develop their understanding about antimicrobials and local antimicrobial usage, and also to build rapport with colleagues from other disciplines.

Communication skills and tools

Effective communication is a critical requirement for meaningful and successful collaboration to occur. The same communication skills needed for effective and meaningful patient interactions are relevant for interprofessional collaboration (Stein-Parbury, 2018). Communication may be verbal or non-verbal, and include skills of active listening, attending, focusing, exploration and mindfulness, along with skills of negotiation and assertiveness (Stein-Parbury, 2018; Webb, 2018). It is important that student and novice nurses are aware of the significance of effective communication skills for successful interprofessional collaboration, and where there is potential for communication to fail. Reflecting on use of one's own communication abilities in practice can be a useful way to develop interprofessional communication skills (Webb, 2018) (see also section on 'Practice Implications').

Negotiation and assertiveness skills are necessary in some clinical practice situations; for example, when there are differences in opinions or priorities. Avoiding situations where assertiveness is needed may be detrimental to meaningful collaboration (Curtis *et al.*, 2011; Wilson *et al.*, 2016). Appropriate use of assertiveness allows individuals to express views in a way that shows respect and is non-judgemental (Curtis *et al.*, 2011; Stein-Parbury, 2018). With use of assertiveness in practice comes a responsibility to maintain good working relationships, and rather than just state a view, which may be perceived as aggressive, to take time to try to understand and acknowledge the perspective of others (Stein-Parbury, 2018). The value of open and direct communication was illustrated in a study exploring interprofessional collaborative practice for medication safety by nursing, medicine and pharmacy students (Wilson *et al.*, 2016). Elements of interprofessional collaborative practice relevant to medication safety happened where professionals 'interacted confidently, contacted each other deliberatively [sic]…asked questions openly…and took action to prevent potentially serious medication errors without feeling constrained' (p. 652). Conversely, displays of passive aggression and other unhelpful communication strategies were seen to be harmful to teamwork and collaboration (Wilson *et al.*, 2016).

Structuring communication enables a shared language to be used amongst health professionals, increasing the likelihood of a shared understanding of information being conveyed, thus minimizing the risk of mistakes occurring due to miscommunication (Foronda *et al.*, 2016; Stein-Parbury, 2018). One example of a structured communication tool is SBAR, which applies a mnemonic for *situation, background, assessment and recommendation*; ISBAR includes a step for *introduction* (Stein-Parbury, 2018). SBAR is particularly useful for communicating urgency in situations where prompt attention to a patient's condition is required (Institute for Healthcare Improvement, http://www.ihi.org/resources/Pages/Tools/SBARToolkit.aspx, accessed 20 November 2019). This includes for sepsis management, which has relevance for AMS, given the importance of accurate clinical assessment and timely administration of antimicrobials (Kleinpell, 2017). However, it is important to appreciate the purpose, so that this means of communication is used appropriately. Wilson *et al.* (2017) developed an online course primarily targeting nurses working in aged care settings, addressing clinical infection management and improving communication using SBAR. Although 100% (n=103) of the respondents completing a pre-post-test survey could correctly define SBAR, only 57% (n=59) could correctly define the *purpose* of SBAR prior to the course (Wilson *et al.*, 2017).

'Scripting' communication may support interprofessional communication. A script for AMS communication was implemented for use in ICU settings in the USA, where an AMS team, in collaboration with nurses, developed 'a change in the script on antimicrobial use that nurses used with physicians' (Jeffs *et al.*, 2018, page 176), which helped to embed AMS within nursing workflow. Sumner *et al.* (2018) recommend and give examples for scripting different collaborative AMS scenarios, including for patient education, allergy assessment and prescriber communication (Sumner *et al.*, 2018).

Applying graded assertiveness techniques aims to assist individuals to escalate concerns about a plan of care through stepped process (Curtis *et al.*, 2011), and may be helpful in situations where nurses are required to negotiate review of antimicrobial prescribing decisions or management plans. Curtis (2011, p. 18) notes that 'gentle cues' may be sufficient to effectively communicate concerns about a patient or about a clinical plan of care to medical staff, prompting doctors to consider a new perspective, or to clarify their perspective (Curtis *et al.*, 2011). Table 7.1 gives an example of how graded assertiveness techniques might be used to raise a concern about antimicrobial management.

Information and communication technology (ICT)

Various health information tools and technologies aim to support interprofessional collaborative practice (Christopherson *et al.*, 2015). Electronic health records, for example, enable patient information, test results, medication records and care plans to be shared between health professionals, thereby enabling all individuals involved in a patient's care to access the same information and data. Many health information technology systems used for AMS incorporate electronic decision support system (eCDSS) functionality, which brings together patient data such

Table 7.1. Example of graded assertiveness and application to AMS. (Adapted from Curtis *et al.*, 2011)

Level	AMS example
Level one: express initial concern with an 'I' statement: *I am concerned about…*	*I'm concerned that this patient has been receiving broad-spectrum antimicrobials for 5 days now. The culture results came back a couple of days ago. I believe she could change to a narrower-spectrum antimicrobial, and this could be given orally. Would you review the patient please?*
Level two: make an enquiry or offer a solution: *Would you like me to...*	*Would you like me to show you the relevant guidelines for antimicrobial prescribing for this condition?*
Level three: ask for an explanation: *It would help me to understand*	*It would help me to know why this antimicrobial choice is being continued for this duration. Can you talk me through this?*
Level four: a definitive challenge demanding a response: *For the safety of the patient you must listen to me*	*Reviewing the patient and the antimicrobial treatment she is receiving, before the next dose is given, is in the best interests of the patient. Please tell me when you will do this?*

as microbiology and results of other diagnostic tests, and information that supports appropriate antimicrobial use, such as guidelines, care protocols and error alerts, to help guide decision-making about patient care and clinical management (ACSQHC, 2018). Having patient care information centrally available helps everyone to be 'on the same page' and have shared understanding about care plans, thus facilitating collaboration and potentially reducing error. Examples of technology with eCDSS include electronic health records, smart phone applications, and electronic prescribing and approval systems (ACSQHC, 2018, page 107).

However, availability of an eCDSS will not always mean health professionals will use the system or that it will be used appropriately (Bardach *et al.*, 2017; ACSQHC, 2018). Moreover, routine use of electronic communication tools may lessen the frequency of face-to-face encounters (Sharma and Klocke, 2014; Bardach *et al.*, 2017), thus removing the opportunities for exploration, clarification and discussion that come with face-to-face communications, and potentially increasing the risk of error through misunderstandings and confusion about patient treatment and care plans (Foronda *et al.*, 2016; Stein-Parbury, 2018). ICT is used to *support* healthcare rather than to manage it. An intentional approach to the design and implementation of information technology systems, which takes into account the context of the health environment and the preparedness of health professionals to practise collaboratively, is important to help ensure use and relevance of the tools in practice (Christopherson *et al.*, 2015). Thus, it is important that nurses be involved in the development and testing of new health information technologies.

Clinical leadership

Interprofessional collaboration has been found to flourish in settings where clinical leaders routinely model collaborative practice skills, but also take action to promote effective communication, trust and respect between health professionals (Robert Wood Johnson Foundation, 2015). Regarding AMS, in a nurse driven AMS programme in intensive care unit settings in South Africa, clinical nurse leaders were seen by doctors, nurses, microbiologists and pharmacists to be effective in communicating and collaborating with individuals from other disciplines, during daily medical rounds, phone discussion, and via face-to-face and telecommunication meetings with microbiology, thus establishing relationships and modelling collaboration skills perceived to be significant to the success of this programme (Rout and Brysiewicz, 2017).

Practice Implications

So, what does all this mean for the student or novice nurse? The road to effective collaboration is not necessarily easy, and for the student or beginning nurse it can appear overwhelming. However, students and new graduates should have confidence in their undergraduate training, and in the fact that they are joining a profession which is grounded in evidence and holds patient-centred care as its core value. New graduate nurses must also remember that, as trusted health professionals, and holding a consistent presence in the patient journey, they are in a pivotal position to influence patient care. Below are some practical suggestions for developing collaborative practice skills, and for engaging with AMS.

- **Maintain a focus on patient-centred care.** Maintaining the nursing focus on patient-centred care is essential. Nurses need to apply their communication skills to help other team members better understand this goal. It is important, too, that patients understand that although many different health professionals are involved in care, there is a shared focus on meeting the care needs of the patient. Given their close relationship with patients, nurses can help educate and empower patients to ask questions and to be involved in decisions about their care, and to clarify the roles of different health professionals so that patients have a better understanding of how teams work together.
- **Reflect on and in practice.** Reflection is a principal aspect of professional practice, and reflection on both positive and negative practice experiences can give valuable insight and learning, especially when documented. Students and beginning practitioners can actively reflect on practice as they begin to engage with AMS, and in doing so identify opportunities to develop their communication abilities and specific AMS skills. Doing this can help individual nurses plan and develop their own toolkit for different types of interprofessional collaboration that may happen in the clinical workplace.
- **Use tools and resources.** Students and new graduates should become familiar with available tools and resources to support them in communicating effectively with other health professionals, which can promote collaborative

practice. This includes guidelines, policies and other communication tools available at the point of care.

- **Practise and model good communication skills.** All nurses, including student and novice nurses, need to consciously practise their communication skills in the real-life clinical environment. Junior nurses may perceive that they do not have influence, but even beginning nurses can role model desired behaviours and encourage use of helpful communication skills. They can also clarify, discuss and question antimicrobial prescribing practices, and encourage and support nursing colleagues to do the same. As part of this, nurses should consider their role and responsibility in reporting behaviours by all colleagues, regardless of discipline, that are not in line with recommended clinical practice, or that are not in keeping with the intent and conditions for good collaborative practice.
- **Engage in networking and learning opportunities.** Taking part in opportunities to network with others from different professional groups, formally and informally, has been suggested as a powerful means of fostering collaboration. Formally, nurses should seek out opportunities to participate in interprofessional meetings, rounds, handovers or interprofessional education activities such as simulation and team-based training, all of which will facilitate collaboration skills and individual confidence in working collaboratively with others.
- **Draw on the strengths of the nursing team (and mentors).** Nursing is focused on teamworking and has at its centre the very values that underpin collaboration: respect, trust and a central focus on patient-centredness. Junior nurses should draw on the strengths of the team to discuss and learn, or raise concerns relating to interprofessional working relationships. Additionally, it can be helpful for novice nurses to identify a mentor in the workplace, and seek to establish a formal mentoring relationship, which can be valuable for learning during the transition from student to registered nurse practice. Many organizations have guidelines for mentoring; therefore, student and novice nurses should seek to identify suitable materials that are available within their local organization or through their nursing body to support them in this.

There are several clinical activities which offer valuable opportunities for developing experience in interprofessional collaboration and for learning and engaging with AMS (Box 7.3).

Box 7.3. Antimicrobial stewardship – opportunities for interprofessional collaboration in AMS.

- Interprofessional rounds, including antimicrobial stewardship team rounds;
- Interprofessional AMS committees and other groups responsible for the; development of guidelines, communication tools and other resources;
- Shared decision-making between health professionals and consumers; about infection management, treatment and care;
- Patient and consumer education;
- Discharge planning and 'Hospital in the Home' programmes;
- Research and quality-improvement activities;
- Audit and feedback sessions;

Antimicrobial Stewardship in Action

Case study 1

Ann is a novice RN who has been working on the surgical ward for a few weeks. Mrs Sook is a 65-year-old woman who has just returned to the ward from the-atre following surgery for bowel carcinoma and has been prescribed six doses of antimicrobial prophylaxis. Mrs Sook has received one dose preoperatively and a second dose in the recovery unit. Ann checks the antimicrobial guidelines for the procedure. She identifies that the guidelines recommend that one dose should be given prior to surgery and that a second dose is given in the recovery unit for longer than expected procedures. No other doses are required. Ann decides she will need to contact the junior surgical team doctor to have the medication reviewed. Ann feels nervous about doing this, as she has only been working on the ward for a few weeks, and fears saying something incorrect during the interaction.

Question 1: Before calling the junior doctor, what information would Ann have needed to gather to help her convey her concerns succinctly to the junior doctor over the phone?

Answer: Sometimes, junior medical staff, or house doctors, may not be fa-miliar with all patients; they may not be aware of their medical history, their overall clinical condition preoperatively, or the reason for surgery. It is often useful to have the medical record at hand when contacting other members of the health team to ask that a patient be reviewed, so any information not col-lected before the call is at hand. Information Ann would need to have so that she was prepared to discuss her concerns includes:

- the patient's full name, date of birth and medical record number;
- the reason for admission, and for surgery;
- whether the patient has any co-morbidities or allergies;
- whether the surgical procedure was uneventful, or if there were complica-tions during surgery, or in the immediate recovery period, and details of the surgical duration;
- the patient's current clinical status, and vital signs;
- details of the medication ordered and the number of doses given; and
- details of current recommended antibiotic guidelines for surgical prophylaxis.

Question 2: Prepare a brief script for the phone call to the junior doctor using the SBAR mnemonic.

Answer: The script should consider the situation, background, assessment and the nurse's recommendation. Some organizations also include an Introduction in the SBAR mnemonic.

Continued

Case study 1 Continued.

Introduction	Hello, my name is Ann; I am one of the registered nurses working on X ward. I am calling you about a patient, Mrs Sook.
Situation and background	Mrs Sook is a 65-year-old woman who returned to the ward from theatre today at midday, following [state name of procedure]. The surgeon has ordered six doses of antimicrobial prophylaxis. Mrs Sook received one dose preoperatively and the second dose while in recovery. She is stable, her vital signs are within expected limits and, at present, she is pain-free. I'm concerned that six doses of antibiotics are unnecessary.
Assessment	Mrs Sook has no other co-morbidities or health problems that have been documented in her initial assessment. Looking at the notes it appears the second dose of antimicrobials was required due to the longer duration of the procedure; but I cannot see any indication documented for continuing the prophylaxis any longer.
Recommendation	I would like you to come and review Mrs Sook and the medication order please.

Question 3: Consider how Ann might develop her knowledge and confidence in communicating with medical staff to help her prepare for similar situations in the future. What are some of the opportunities available for her as part of her routine practice?

Answer: Ann could develop her knowledge and confidence in several ways:

- Exploring education and training opportunities focused on developing interprofessional communication and collaboration skills, as well as those focused on AMS; including interprofessional meetings, rounds and handovers, and interprofessional education activities such as simulation and team-based training, which are available to health professionals from different disciplines.
- Reflection on both positive and negative practice experiences to identify opportunities to develop her communication abilities and specific AMS skills. Use of a specific reflective practice tool can help nurses to structure their reflections, and to determine actions to develop their skills and set realistic personal goals.
- Learning about and using available clinical tools and resources, such as guidelines, policies and communication tools available at the point of care that help to support day-to-day clinical practice and collaborative care.
- Draw on other members of the nursing team to discuss and learn about collaborating effectively with other health professionals, and to learn more

Continued

Case study 1 Continued.

about good antimicrobial management (senior or more experienced nurses are often well placed to advise and to facilitate education of nurses about infection and antimicrobial management, and about interprofessional working relationships).

- Explore options for establishing a mentoring relationship with a more senior or experienced nurse.
- Contribute to resource development, or to policy revision, or become involved in piloting new clinical resources or tools.

Case study 2

Harry and Vidya are undertaking a medication round. Harry is an experienced nurse, while this is Vidya's first clinical placement as a registered nurse. Harry checks the medication chart and the medical record for the next patient. After reading the electronic health record, Harry tells Vidya that Mr Sansa's antimicrobial treatment should be changed from the intravenous to the oral route. After the medication round is complete, Harry speaks with the junior medical officer to discuss the potential change from intravenous to oral antibiotics, and the junior doctor confirms that this change can be done. The junior doctor also advises Harry that Mr Sansa should be ready for discharge the next day, and Harry asks Vidya to let Mr Sansa know this. Vidya speaks to Mr Sansa, saying, 'I'm sure you'll be happy to know that you will probably be going home tomorrow, Mr Sansa'. Mr Sansa expresses some concern, saying he still feels weak and unwell, and that he is nervous about stopping IV antibiotics. He states that he feels he really should stay in hospital and continue to receive some 'stronger' antibiotics through the 'drip'. He asks for Vidya's opinion about the change from IV to oral antibiotics. Vidya feels she is not adequately equipped to answer Mr Sansa's question or to discuss his concerns.

Question 1: Why is it important that all members of an interprofessional team, including the patient, understand the rationale for a patient's individual antibiotic management plan?

Answer: Antimicrobial stewardship is about all aspects of antimicrobial management – from clinical assessment, diagnosis, prescription, dispensing, administration and ongoing monitoring. Thus, responsibility for AMS is something that should be shared between all members of the health team; it is not something that is owned by only one individual health professional. As with any clinical practice, to deliver safe care health professionals need to understand why treatment is being delivered, and the expected outcome, so that they know what they are assessing and monitoring for, and recognize early if the patient is not improving or progressing as expected. The patient also needs to know what to expect. Understanding the rationale, and the plan, helps team members to know how they contribute to the plan of care and what their part in the interprofessional

Continued

Case study 2 Continued.

team is. It minimizes the risk of confusion among team members, errors in care delivery, and helps ensure health professionals are equipped to provide consistent information to patients and other health professionals.

Question 2: How could Vidya respond to Mr Sansa's concerns? Which members of the interprofessional team could Vidya call on for support and advice?

Answer:

- A priority in this case is for Vidya to first acknowledge Mr Sansa's concerns about not feeling ready to go home, and about his treatment. Trust is important in interprofessional relationships, and this includes trust between the health professional (nurse) and the patient. Although Vidya is not feeling confident about responding, she could repeat and reframe what Mr Sansa has said and expressed, to confirm she understands his concerns.
- Vidya could then let Mr Sansa know that although she knows that many antibiotics are just as effective in treating infections as those given via the intravenous route, she has not spoken directly with the medical staff about this specific change, and that she will seek some more detail from her colleagues about the plan of care. To ensure trust is maintained, it would be important that Vidya follow up on this as soon as practicable.
- Vidya could then consider speaking with Harry to discuss her concerns about communicating with the patient, and to clarify if Harry is able to provide more details, following his phone conversation with the medical officer. Harry, as the senior nurse, may also be able to explore further with Mr Sansa his concerns about being unwell, and about changing the route of antimicrobial therapy. He may also be able to let Mr Sansa know more about the benefits of making the transition from intravenous to oral therapy, and the safety and efficacy of this specific to the antibiotics that Mr Sansa has been prescribed.
- If Harry is not available, Vidya could contact the junior medical officer and request that they come and speak with Mr Sansa. However, before doing this, it would be wise for Vidya to have explored with Mr Sansa the basis of his comments about feeling weak and unwell, and his concerns about stopping IV antibiotics, so that when she speaks with the junior doctor she can convey the necessary information, and advocate for Mr Sansa's involvement in his care planning. If Vidya is uncertain about doing this, and Harry is not available, another more senior nursing colleague would likely be able to support Vidya, through reviewing the medical record together and discussing the next steps for supporting Mr Sansa.

Later that week, Vidya plans to attend an education session on the ward, which is focused on providing feedback to staff about hospital antimicrobial prescribing practices. Vidya asks one of the other nurses if she is attending. The other nurse tells Vidya: 'Apparently, there is some sort of study going on about antibiotics being given for too long. It is only about the doctors and their prescribing. No nurses are involved, and I can't prescribe anyway, so I'm not going to the

Continued

Case study 2 Continued.

---

presentation'. Vidya feels confused about whether to attend the education session, and wonders if it will be of relevance to her after all.

Question 3: Why is it important that nurses are aware of local antimicrobial usage and resistance surveillance activities, and associated findings of these activities?

Answer:
Surveillance of antimicrobial usage and resistance, followed by feedback on results of this surveillance to health professionals, is a core AMS activity. Information obtained from surveillance activities guides practice and clinical decision-making, and underpins local AMS policy development. Having information about the purpose and findings of local surveillance activities helps nurses understand where the priorities for AMS are, raising their awareness about which antimicrobials are of concern in terms of their overuse, or misuse, or in driving local antimicrobial resistant infections. Having knowledge and understanding of the priorities for improvement in AMS can also help the team to develop a shared focus for improvement. Having a shared focus on a specific goal of care and knowing how an individual health professional contributes to this goal can encourage collaboration about antimicrobial usage.

Question 4: In what ways might attendance at an interprofessional education session focused on antimicrobial surveillance help to foster better interprofessional understanding, and help nurses develop their skills in interprofessional collaborative practice?

Answer:
Interprofessional education (IPE) may help to develop an individual's communication and teamworking skills and help to improve understanding among health professionals about the roles and values of other health disciplines. During an education and feedback session focused on results of local antimicrobial usage surveillance, participants will likely have an opportunity to listen to other perspectives, and to discuss the findings and potential activities to improve. This can help to ensure everyone understands the priorities and goals of the local stewardship programme, thereby giving a shared focus towards which team members are working. Joining with other health professionals during educational activities may assist individuals to better understand the roles of team members from other disciplines, and in doing so help nurses develop their capacity for interprofessional collaboration.

Key points

- Antimicrobial stewardship requires effective collaboration between health professionals from different disciplines.
- Understanding the roles and perspectives of different members of the interprofessional team is important for effective interprofessional collaboration.

- Nurses are keen to collaborate, but this keenness may be impacted by internal and external factors, including knowledge, confidence, role clarity, and team and organizational culture.
- Strategies, tools and resources exist to encourage and support interprofessional collaboration in antimicrobial stewardship.
- Nurses can engage in specific activities to develop and promote interprofessional collaborative practice as part of antimicrobial stewardship.

Suggested Further Reading

Castro-Sánchez, E., Gilchrist, M., McEwen, J., Smith, M., Kennedy, H. and Holmes, A. (2017) Antimicrobial stewardship: widening the collaborative approach. *Journal of Antimicrobial Stewardship*, March.

Charani, E., Castro-Sánchez, E., Sevdalis, N., Kyratsis, Y., Drumright, L., Shah, N. and Holmes, A. (2013) Understanding the determinants of antimicrobial prescribing within hospitals: the role of 'prescribing etiquette'. *Clinical Infectious Diseases* 57, 188–196.

Ewers, T., Knobloch, M.J. and Safdar, N. (2017) Antimicrobial stewardship: the role of the patient. *Current Treatment Options in Infectious Diseases* 9, 92–103.

Petri, L. (2010) Concept analysis of interdisciplinary collaboration. *Nursing Forum* 45, 73–82.

Stein-Parbury, J. (2018) *Patient and Person Interpersonal Skills in Nursing*. Elsevier.

Wilson, A. J., Palmer, L., Levett-Jones, T., Gilligan, C. and Ooutram, S. (2016) Interprofessional collaborative practice for medication safety: nursing, pharmacy, and medical graduates' experiences and perspectives. *Journal of Interprofessional Care* 30, 649–654.

References

American Nursing Association and Centers for Disease Control and Prevention (2017) Redefining the antibiotic stewardship team: recommendations from the American Nurses Association/Centers for Disease Control and Prevention Workgroup 19 on the role of registered nurses in hospital antibiotic stewardship practices. *Nursing World*.

ACSQHC (Australian Commission on Safety and Quality in Health Care) (2018) *Antimicrobial Stewardship in Australian Health Care*. ACSQHC, Sydney, Australia.

Bardach, S.H., Real, K. and Bardach, D.R. (2017) Perspectives of healthcare practitioners: an exploration of interprofessional communication using electronic medical records. *Journal of Interprofessional Care* 31, 300–306.

Barlam, T.F., Cosgrove, S.E., Abbo, L.M., MacDougall, C., Schuetz, A.N., *et al.* (2016) Implementing an antibiotic stewardship program: guidelines by the Infectious Diseases Society of America and the Society for Healthcare Epidemiology of America. *Clinical Infectious Diseases* 62, e51–e77.

Braithwaite, J., Herkes, J., Ludlow, K., Testa, L. and Lamprell, G. (2017) Association between organisational and workplace cultures, and patient outcomes: systematic review. *BMJ Open* 7, e017708.

Broom, A., Broom, J., Kirby, E. and Scambler, G. (2016) Nurses as antibiotic brokers: institutionalized praxis in the hospital. *Qualitative Health Research* 30, 30.

Carter, E.J., Greendyke, W.G., Furuya, E.Y., Srinivasan, A., Shelley, A.N., *et al.* (2018) Exploring the nurses' role in antibiotic stewardship: a multisite qualitative study of nurses and infection preventionists. *American Journal of Infection Control* 46, 492–497.

Castro-Sánchez, E., Gilchrist, M., McEwen, J., Smith, M., Kennedy, H. and Holmes, A. (2017) Antimicrobial stewardship: widening the collaborative approach. *Journal of Antimicrobial Stewardship*, March.

Charani, E., Edwards, R., Sevdalis, N., Alexandrou, B., Sibley, E., *et al.* (2011) Behavior change strategies to influence antimicrobial prescribing in acute care: a systematic review. *Clinical Infectious Diseases* 53, 651–652.

Charani, E., Cástro-Sanchez, E., Sevdalis, N., Kyratsis, Y., Drumright, L., Shah, N. and Holmes, A. (2013) Understanding the determinants of antimicrobial prescribing within hospitals: the role of 'prescribing etiquette'. *Clinical Infectious Diseases* 57, 188–196.

Charani, E., Smith, I., Skodvin, B., Perozzello, A., Lucet, J.-C., *et al.* (2019) Investigating the cultural and contextual determinants of antimicrobial stewardship programmes across low-, middle- and high-income countries – a qualitative study. *PLOS ONE* 14, e0209847.

Christopherson, T.A., Troseth, M.R. and Clingerman, E.M. (2015) Informatics-enabled interprofessional education and collaborative practice: a framework-driven approach. *Journal of Interprofessional Education & Practice* 1, 10–15.

Cosgrove, S.E., Hermsen, E.D., Rybak, M.J., File, T.M., Parker, S.K. and Barlam, T.F. (2014) Guidance for the knowledge and skills required for antimicrobial stewardship leaders. *Infection and Control in Hospital Epidemiology* 35, 1444–1451.

Cotta, M.O., Robertson, M.S., Marshall, C., Thursky, K.A., Liew, D. and Buising, K.L. (2015) Implementing antimicrobial stewardship in the Australian private hospital system: a qualitative study. *Australian Health Review* 39, 315–322.

Courtenay, M., Lim, R., Cástro-Sanchez, E., Deslandes, R., Hodson, K., *et al.* (2018) Development of consensus-based national antimicrobial stewardship competencies for UK undergraduate healthcare professional education. *The Journal of Hospital Infection* 100, 245–256.

Curtis, K., Tzannes, A. and Rudge, T. (2011) How to talk to doctors – a guide for effective communication. *International Nursing Review* 58, 13–20.

D'Amour, D., Ferrada-Videla, M., San Martin Rodriguez, L. and Beaulieu, M.-D. (2005) The conceptual basis for interprofessional collaboration: core concepts and theoretical frameworks. *Journal of Interprofessional Care* 19, 116–131.

Davey, P. (2015) The 2015 Garrod Lecture: Why is improvement difficult? *Journal of Antimicrobial Chemotherapy* 70, 2931–2944.

Dellit, T.H., Owens, R.C., McGowan, J.E., Gerding, D.N., Weinstein, R.A., *et al.* (2007) Infectious Diseases Society of America and the Society for Healthcare Epidemiology of America guidelines for developing an institutional program to enhance antimicrobial stewardship. *Clinical Infectious Diseases* 44, 159–177.

Edwards, R., Drumright, L.N., Kiernan, M. and Holmes, A. (2011) Covering more territory to fight resistance: considering nurses' role in antimicrobial stewardship. *Journal of Infection Prevention* 12, 6–10.

Ewers, T., Knobloch, M.J. and Safdar, N. (2017) Antimicrobial stewardship: the role of the patient. *Current Treatment Options in Infectious Diseases* 9, 92–103.

Fisher, C.C., Cox, V.C., Gorman, S.K., Lesko, N., Holdsworth, K., Delaney, N. and McKenna, C. (2018) A theory-informed assessment of the barriers and facilitators to nurse-driven antimicrobial stewardship. *American Journal of Infection Control* 46, 1365–1369.

Foronda, C., MacWilliams, B. and McArthur, E. (2016) Interprofessional communication in healthcare: an integrative review. *Nurse Education in Practice* 19, 36–40.

Gillespie, E., Rodrigues, A., Wright, L., Williams, N. and Stuart, R.L. (2013) Improving antibiotic stewardship by involving nurses. *American Journal of Infection Control* 41, 365–367.

Gonzalo, J., Kuperman, E, Lehman, E. and Haidet, P. (2014) Bedside Interprofessional rounds. *Journal of Hospital Medicine* 10, 646–651.

Hall, P. (2005) Interprofessional teamwork: professional cultures as barriers. *Journal of Interprofessional Care* 19 (Suppl 1), 188–196.

Hughes, R.G. (2008) Nurses at the 'sharp end' of patient care. In: Hughes, R.G. (ed.) *Patient Safety and Quality: An Evidence-Based Handbook for Nurses.* Agency for Healthcare Research and Quality (US), Rockville, Maryland.

Institute for Healthcare Improvement (IHI). *SBAR Tool: Situation-Background-Assessment-Recommendation.* Cambridge, Massachusetts. Available at: http://www.ihi.org/resources/Pages/Tools/SBARToolkit.aspx (accessed May 2019).

Jeffs, L., Law, M.P., Zahradnik, M., Steinberg, M., Maione, M. (2018) Engaging nurses in optimizing antimicrobial use in ICUs: a qualitative study. *Journal of Nursing Care Quality* 33, 173–179.

Kleinpell, R. (2017) Promoting early identification of sepsis in hospitalized patients with nurse-led protocols. *Critical Care* 21, 10.

Levett-Jones, T., Burdett, T., Chow, Y.L., Jönsson, L., Lasater, K., Matthews, L.R., *et al.* (2018) Case studies of interprofessional education initiatives from five countries. *Journal of Nursing Scholarship* 50, 324–332.

Manley, K., Sanders, K., Cardiff, S. and Webster, J. (2011) Effective workplace culture: the attributes enabling factors and consequences of a new concept. *International Practice Development Journal* 1(2), art. 1. Available at: https://www.fons.org/library/journal/volume1-issue2/article1 (accessed 30 September 2019).

McCallin, A. (2001) Interdisciplinary practice – a matter of teamwork: an integrated literature review. *Journal of Clinical Nursing* 10, 419–428.

Messina, A.P., Van den Bergh, D. and Goff, D.A. (2015) Antimicrobial stewardship with pharmacist intervention improves timeliness of antimicrobials across thirty-three hospitals in South Africa. *Infectious Diseases and Therapy* 4, 5–14.

Monsees, E., Popejoy, L., Jackson, M.A., Lee, B. and Goldman, J. (2018) Integrating staff nurses in antibiotic stewardship: opportunities and barriers. *American Journal of Infection Control* 46, 737–742.

Olans, R.D., Nicholas, P.K., Hanley, D. and DeMaria, A. (2015) Defining a role for nursing education in staff nurse participation in antimicrobial stewardship. *Journal of Continuing Education in Nursing* 46, 318–321.

Olans, R.N., Olans, R.D. and DeMaria, A. (2016) The critical role of the staff nurse in antimicrobial stewardship: unrecognized, but already there. *Clinical Infectious Diseases* 62, 84–89.

Petri, L. (2010) Concept analysis of interdisciplinary collaboration. *Nursing Forum* 45, 73–82.

PHE (Public Health England) (2011) *Start Smart – then Focus. Antimicrobial Stewardship Toolkit for English Hospitals*. Public Health England, London.

Raphael-Grimm, T. (2014) Art of communication in nursing and health care: an Interdisciplinary Approach. Springer Publishing Company, New York.

Reeves, S. (2012) The rise and rise of interprofessional competence. *Journal of Interprofessional Care* 26, 253–255.

Reeves, S., Perrier, L., Goldman, J., Freeth, D. and Zwarenstein, M. (2013) Interprofessional education: effects on professional practice and healthcare outcomes. *Cochrane Database of Systematic Reviews*.

Reeves, S., Pelone, F., Harrison, R., Goldman, J. and Zwarenstein, M. (2017) Interprofessional collaboration to improve professional practice and health-care outcomes. *Cochrane Database of Systematic Reviews* 6, Cd000072.

Reeves, S., Xyrlchis, A. and Zwarenstein, M. (2018) Teamwork, collaboration, co-ordination, and networking: why we need to distinguish between different types of interprofessional practice. *Journal of Interprofessional Care* 32, 1–3.

Rixon, S., Braaf, S., Williams, A., Liew, D. and Manias, E. (2015) Pharmacists' inter-professional communication about medications in specialty hospital settings. *Health Communication* 30, 1065–1075.

Robert Wood Johnson Foundation (2015) Lessons from the field: promising inter-professional collaboration practices. White paper. Robert Wood Johnson Foundation.

Rout, J. and Brysiewicz, P. (2017) Exploring the role of the ICU nurse in the anti-microbial stewardship team at a private hospital in KwaZulu-Natal, South Africa. *Southern African Journal of Critical Care* 33, 46–50.

Sharma, U. and Klocke, D. (2014) Attitudes of nursing staff toward interprofessional in-patient-centered rounding. *Journal of Interprofessional Care* 28, 475–477.

Stein-Parbury, J. (2018) *Patient & Person Interpersonal Skills in Nursing*. Elsevier.

Stuart, R.L., Orr, E., Kotsanas, D. and Gillespie, E. (2015) A nurse-led antimicro-bial stewardship intervention in two residential aged care facilities. *Healthcare Infection* 20, 4–6.

Sumner, S., Forsyth, S., Colette-Merrill, K., Taylor, C., Vento, T., Veillette, J. and Webb, B. (2018) Antibiotic stewardship: the role of clinical nurses and nurse educators. *Nurse Education Today* 60, 157–160.

Tang, C.J., Chan, S.W., Zhou, W.T. and Liaw, S.Y. (2013) Collaboration between hospital physicians and nurses: an integrated literature review. *International Nursing Review* 60, 291–302.

Thistlethwaite, J.E., Forman, D., Matthews, L.R., Rogers, G.D., Steketee, C. and Yassine, T. (2014) Competencies and frameworks in interprofessional educa-tion: a comparative analysis. *Academic Medicine* 89, 869–875.

Tromp, M., Hulscher, M., Bleeker-Rovers, C.P., Peters, L., Van den Berg, D.T.N.A., *et al*. (2010) The role of nurses in the recognition and treatment of patients with sepsis in the emergency department: a prospective before-and-after intervention study. *International Journal of Nursing Studies* 47, 1464–1473.

Webb, L. (2018) Exploring the characteristics of effective communicators in healthcare. *Nursing Standard* 33, 47.

Wilson, A.J., Palmer, L., Levett-Jones, T., Gilligan, C. and Outram, S. (2016) Interprofessional collaborative practice for medication safety: nursing, pharmacy, and medical graduates' experiences and perspectives. *Journal of Interprofessional Care* 30, 649–654.

Wilson, B.M., Shick, S., Carter, R.R., Heath, B., Higgins, P.A., *et al.* (2017) An on-line course improves nurses' awareness of their role as antimicrobial stewards in nursing homes. *American Journal of Infection Control* 45, 466–470.

WHO (World Health Organization) (2010) *Framework for Action on Interprofessional Education & Collaborative Practice*. World Health Orgnanization, Geneva, Switzerland. Available at: https://www.who.int/hrh/resources/framework_action/en/ (accessed May 2019).

WHO (2016) *Global Strategic Directions for Strengthening Nursing and Midwifery 2016–2020*. World Health Organization, Geneva, Switzerland. Available at: https://www.who.int/hrh/nursing_midwifery/global-strategic-midwifery2016-2020.pdf (accessed 7 January 2020).

Zabarsky, T.F., Sethi, A.K. and Donskey, C.J. (2008) Sustained reduction in inappropriate treatment of asymptomatic bacteriuria in a long-term care facility through an educational intervention. *American Journal of Infection Control* 36, 476–480.

Leading and Supporting Antimicrobial Stewardship

8

Rose Gallagher[1], Rita Olans[2], Susie Singleton[3] and Joanne Bosanquet[4]

[1]*Professional Lead Infection Prevention and Control/AMR & Sustainability Lead, Royal College of Nursing, London;* [2]*Assistant Professor, School of Nursing, MGH Institute of Health Professions, Boston, USA;* [3]*Consultant Nurse, Health Protection & Infection Prevention and Control, Public Health England, London;* [4]*Chief Executive, Foundation of Nursing Studies, London*

Objective: To provide an overview of the role and influence of nursing leadership and the practical implementation of antimicrobial stewardship (AMS) in nursing practice.

International and National AMS Strategy/Policy

The antimicrobial resistance (AMR) global crisis has been described as a 'burning platform' that will affect every human being if we do not strive to reduce and control it (Bosanquet and Singleton, 2014). As the same or similar antimicrobials can be used to treat infectious diseases in humans and animals, with resistant bacteria arising and spreading from one to the other or the environment, and from country to country, tackling such a huge challenge requires a holistic and multisectoral (or 'One Health') approach (WHO, 2017).

Nurses and midwives are a vital resource to help address this challenge. The International Council of Nurses (ICN) defines nursing as an integral part of the healthcare system, encompassing the promotion of health, prevention of illness and care of physically ill, mentally ill and disabled people of all ages, in all healthcare and other community settings (ICN, 2019). Nursing is a profession that cares for individuals within the context of the family and the community, a central focus of its contribution to AMS (Clarke and Bleich, 2018). To be successful in specifically addressing AMR, nurses need to be aware of how they can contribute to AMS programmes and how they can exert leadership to increase public education and awareness of AMR.

Recognizing the nursing contribution to current global health challenges such as AMR, it is important to first acknowledge the relative late arrival of the nursing profession to global policy development and implementation. Despite accounting for nearly 50% of the healthcare workforce worldwide (WHO, 2018), a Chief Nursing Officer was only appointed at the WHO in 2017. This is perhaps symptomatic of an oversight of nursing globally, and a lack of recognition with regards to how nursing can contribute to health, and its associated human and economic benefits. For AMR, such dearth of recognition of nursing's contribution has also been identified (Courtenay *et al.*, 2019), with a need for greater interprofessional working at all levels recognized (Davies, 2019).

Strategic Leadership

Leadership in nursing has, historically, been associated with formal leadership programmes and nurses working in executive roles. However, current global nursing and midwifery leaders have been instrumental in positioning the profession at the heart of global health and care policy and delivery, moving beyond historically routine but limited position statements (ICN, 2017) to impactful campaigns such as 'Nursing Now' and the 'Triple Impact Report' (All-Party Parliamentary Group (APPG) on Global Health, 2016). This has both raised the profile and contribution of nursing, and quantified it in health policy language that resonates with political, business and global health leaders. With the emergence of AMS programmes, there is an opportunity for clinical nursing leadership. A call to action to challenge the 'stereotype of nursing', to raise its profile and engage better on a multi-professional level is necessary to create meaningful AMS strategies that move beyond the medical and pharmaceutical dogma to one of broader, care and public health actions where nurses and midwives excel. Explicit literature on the role of nursing and midwifery in AMS is crucial to the development of nursing knowledge and skills in this area. For example, the American Nurses Association/ Centers for Disease Control and Prevention (ANA/CDC) White Paper (2017) clearly describes the nurse's role in hospital AMS practices, and the wide-ranging and often unrecognized roles that nurses play in AMS have been clearly aligned with good nursing practice (Olans *et al.*, 2016; Castro-Sánchez *et al.*, 2017). These roles exceed the traditional focus on the prevention of infection and recognize other practical nursing skills that amplify the contribution of other professionals such as pharmacists and doctors.

Strategic nursing leaders concerned with influencing local, national and global health policy and advocating engagement with AMS on behalf of the profession should:

- be visible;
- work closely with other nurses/midwives in all roles and care settings to unify nursing policy and practice, to maximize the impact of stewardship activities; and

- ensure nursing has 'a seat at the table' for all interprofessional policy and practice development (see Chapter 7), an often-overlooked capacity due to the ubiquity and thus invisibility of the unique contribution of nursing in organizing work (Allen, 2018).

Such engagement and visibility would require leaders to adopt behaviours and attitudes similar, for example, to these suggested by the NHS Leadership Academy (2019):

- Inspire shared purpose;
- Lead with care;
- Evaluate information;
- Connect services;
- Share a vision;
- Engage the team;
- Hold others to account;
- Develop capability;
- Influence for results.

Finally, nursing and midwifery leaders should ensure that, regardless of role or position, nurses and midwives understand that they have a contribution to stewardship, and that such contribution is not only circumscribed to clinical skills but also encompasses the leadership facet that exists in each and every position to influence clinical practice (as outlined in Chapter 2) and use of antibiotics (Chapter 4). Examples of the different contributions to AMS nurses can make, across a variety of settings and roles, are described below (Table 8.1). Whilst not exhaustive, this list showcases some of the activities that contribute to the broad elements of AMS that extend to clinical practice.

Implementation of AMS in Clinical Practice

Initial AMS programmes did not include nurses among their core human resources (Shlaes *et al.*, 1997). However, more recent policy documents provide some clarity about the contribution nurses can make to AMS – for example, those working in specialist fields such as infection prevention and control – and the role nurses play with regards to patient safety (Nathwani, 2012). However, examination of core daily nursing functions makes obvious the activities that nurses perform that could be embedded in stewardship and that reflect principles of stewardship practice (Olans *et al.*, 2016). Such participation, present but perhaps not explicitly recognized, has been highlighted by professional leadership organizations such as the ANA/CDC, Agency for Healthcare Research and Quality (AHRQ), The Joint Commission (TJC), Centers for Disease Control and Prevention (CDC) and Royal College of Nursing (RCN). The contribution of nursing practice to AMS should not be limited to the hospital setting (Olans *et al.*, 2019).

Table 8.1. Examples of the contribution of different nursing roles to AMS. (Adapted from ANA/CDC, 2017)

Role	Examples in practice			
Clinically based nurses/midwives	Appropriate triage and isolation of patients	Clinical assessment and appropriate specimen collection	Reviews antibiotic prescriptions dose/time for accuracy, checks allergy status, and administers and records the antibiotic administration	Patient education
Infection Prevention and Control (IPC) specialist	Provides specialist advice on the management of infection	Manages outbreaks of infection	Advises on the design of the built environment to reduce the risk of infection	Healthcare worker education
Public health	Health promotion	Travel health advice	Reduction of health inequalities	Public education
Health visitor/school nurse	Seeks consent for and administers vaccination	Assesses the development of children for maximum health	Health screening	Child/parent education on vaccination
Health protection	Surveillance, promotion and gathering data to build the evidence base	Provides advice on biological, chemical and radiological risks	Supports the management of outbreaks	Contributes to the education of healthcare workers and partner organizations including academia

As seen in other closely related fields such as hand hygiene and infection prevention and control, successful implementation of interventions, and specifically those aiming to improve care quality, benefit from the explicit endorsement and buy-in from senior leaders. Nurses in such positions should therefore proactively engage with other leaders during any policy development process to identify how and where they should demonstrate their support to stewardship interventions and actions, as well as clarifying the nursing contribution within them. Such nursing leadership engagement should include the internal and external advocacy for the inclusion of nurses among stewardship programmes (Castro-Sánchez *et al.*, 2018).

Nursing leaders can have further impact in AMS by promoting and supporting the implementation and evaluation of nursing competencies in AMS (Courtenay *et al.*, 2019). The implementation of these or other competencies will vary in different workplace settings, in different countries, and with different resources (Goff *et al.*, 2017), and local idiosyncrasies in terms of professional roles and scope of practice, legislation or role traditions will have to be observed. However, irrespective of location or workplace site (Jump *et al.*, 2017; Rout and Brysiewicz, 2017) the implementation of stewardship principles and achievement of stewardship goals entail the involvement of nurses in the stewardship process in all stages of patient care (Castro-Sánchez, 2018). In order to achieve this, a new understanding is required of the optimal appropriate use of antibiotics in the 21st century (Spellberg, 2016). In turn, this will necessitate the commitment of nursing leadership (in academia, professional bodies and healthcare facilities) to champion this endeavour across both the educational and professional nursing journey, from undergraduate health education to a widespread approach of the understanding of antibiotic use throughout society (WHO, 2017) and therefore to opportunities for nursing support and involvement.

Due to its infancy and the importance of a transformational approach, nursing leadership within AMS should be viewed within a practice development framework (PDF). A PDF encompasses person-centred practices and approaches to leadership. This approach is vital to building relationships with health and care sector organizations, as well as individuals, families and communities.

Case Study 1: Multi-drug Resistant Urinary Tract Infection

Background

Urinary tract infections (UTIs) are known to be the most common cause of opportunistic infections, often resulting from people's own bacterial flora, are often over-diagnosed when colonization is present, and drive a large proportion of antibiotic prescribing in hospital and acute settings (Collins, 2008). They are a major driver for antibiotic use globally, as well as one of the main causes of sepsis (RCN, 2016).

This case study describes a prolonged series of events with implications for the patient, health and care services and the healthcare environment.

Continued

Scenario development

Martha, a 70-year-old woman with a past medical history of dementia and presumed recurrent UTI, had been an in-patient in a care home for 12 months. She sustained a fall that necessitated admission to hospital and had a long history of recurrent UTI requiring hospital care. She was well known to the local acute hospital and had previously been found to carry a Carbapenemase Resistant Organism (CRO) in her urine, with IPC precautions put in place to manage this on each admission episode.

Measuring impact

Martha had a long history of treatment for presumed recurrent urinary tract infections. Care home staff frequently performed urinalyses by dipstick if her urine was dark or 'smelly' and antibiotics were prescribed when the dipstick registered positive for esterase (white blood cells) and nitrites (bacteria). Nursing progress notes never indicated the presence of clinical symptoms of infection such as fever or flank pain. Urine cultures were sent only occasionally.

After years of treatment with antibiotics for suspected UTIs, a urine culture obtained from an indwelling bladder catheter grew a *Klebsiella pneumoniae* that was resistant to all antibiotics tested including carbapenems. This was subsequently found to be a highly resistant CRO.

Martha's care home refused to accept her return after her fall stating they were unable to care for an individual carrying this resistant organism. They stated that the room she occupied did not have en suite facilities and the care home residents shared communal areas including a bathroom. Martha was kept on a general medical ward as an interim measure pending placement. From an acute hospital perspective this was disruptive as an acute hospital bed remained unavailable, and Martha became increasingly distressed as a result of not returning to the familiar surroundings of her care home. Together with the care home staff, a support meeting was convened to assess risks and review the current situation. Members of the incident team included: consultant microbiologist, consultants in public health, senior nurses in health protection, IPC nurses, matrons, social services, care home managers and commissioning personnel.

Discussions provided opportunities to gain further detail and enabled a dynamic risk assessment to be made. The care home facility was finally mollified after a plan was agreed to screen all current residents in the home for evidence of antibiotic-resistant *Klebsiella* infection. No other cases of antibiotic-resistant *Klebsiella* infection were identified, which reassured the care home staff and facilitated Martha's return.

Education was provided to care home staff on IPC measures, specifically the prevention of unnecessary indwelling catheter urine dipstick testing, the appropriate disposal of urine and body fluids, hand hygiene and the correct use of personal protective equipment (PPE) such as gloves and aprons. This education inspired two home staff to become champions to sustain local learning and practice.

Continued

Case Study 1: Continued.

The role of leadership in Martha's care

Nursing leadership was able to be exercised to deliver a successful outcome for all those involved in this scenario. As holistic observers, nurses in the hospital were able to identify the wider implications of delayed discharge and to use their skills and clinical knowledge to both manage the situation and prevent further episodes from occurring. These can be summarized within the nine leadership dimensions (NHS Leadership Academy, 2019) (Table 8.2).

Table 8.2. Nursing leadership in the context of multi-drug resistant urinary tract infections.

Leadership dimension	Application in practice
Inspiring shared purpose Improving catheter care and IPC practice locally for the benefit of all.	The senior IPC nurse negotiating Martha's return must inspire the care home manager to invest in changes to embedded cultures of practice for patients with urinary catheters, and support care workers to understand how this can benefit the wider system and spread of antibiotic resistance locally.
Evaluating information The IPC nurse must use data to initiate and sustain continuous improvements in the care of patients with urinary catheters in the home.	Using laboratory and surveillance data to monitor and reassure staff of the consequences of changes to practice (e.g. not using dipsticks, culturing only when symptoms present) is instrumental to successful change. Over time, the care home is able to see the impact of the changes they have made with significant reductions in the number of patients receiving antibiotics.
Engaging the team Nurses working in different roles collaborate to support staff in the care home to improve patient care and raise the profile of AMS. This extends beyond IPC to include nutrition and hydration and the physical/mental impediments to this.	Nurses involved in patient discharge, community health services and dementia care work together to support the care home manager. Fostering a supportive relationship to ensure cooperation to safeguard patient safety and bring mutual benefit for both organizations are the primary goals. Investment in nurse-led education for staff results in a shared common understanding and purpose to a holistic approach to care and the prevention of infection.

Continued

Case Study 1: Continued.	
Table 8.2. Continued.	
Leadership dimension	Application in practice
Developing capability Nurses act as role models and inspire others to develop and take on new roles.	As a result of receiving professional support and education, two of the nurses in the care home express an interest in acting as link nurses. They join the link nurse programme at the local hospital, developing their knowledge and influencing skills to take back to the care home setting. IPC audits over the following 12 months show improvements in compliance as well as reduced admissions to the local hospital for urinary tract infections.

Case Study 2: Impact of Vaccination on Antimicrobial Resistance

Background

Vaccination is an intervention which has saved the greatest number of lives globally and reduced the burden of disease (Greenwood, 2014). Vaccinations, whether for viral or bacterial infections, have an impact on antimicrobial resistance in various ways including:

- a reduction in the numbers of circulating pathogens;
- reduction of febrile illnesses, so preventing unnecessary antibiotic prescribing;
- prevention of secondary bacterial infections following a viral illness; and
- protecting populations who are not vaccinated through herd immunity.

Scenario development

Max is an 83-year-old man who fell at home and was admitted to hospital and diagnosed with a fracture of the wrist. Prior to this admission, he was living at home with limited mobility. He had frequent visits by his daughter and grandchildren, including the newest granddaughter who was 2 months old. Upon admission to the emergency room Max was asked if he had received his influenza and pneumococcal vaccines. He declined the proffered vaccinations due to concerns over what he had heard in the media about vaccines.

Max was treated for his fractured wrist and discharged. Two days later, Max began to feel generally unwell and developed a fever, cough, body aches, fatigue and headache. His daughter, concerned about Max's symptoms, came to visit him with her children and their infant child. Max developed pneumonia, necessitating hospitalization. He was diagnosed with pneumococcal pneumonia secondary to acute influenza A. His illness contributed to further debilitation requiring permanent nursing home residence.

Continued

Subsequently, his 2-month-old granddaughter developed first an ear infection then pneumococcal bacteremia and meningitis, requiring a prolonged admission to the pediatric intensive care unit. The infant was hospitalized for several weeks and was ultimately sent home well but with residual hearing loss.

Measuring impact

Acute influenza A infection has well-described secondary bacterial infection sequelae with both *Streptococcus pneumoniae* (occasionally penicillin-resistant), as well as Staphylococcal (including Meticillin-resistant *Staphylococcus aureus* [MRSA]) infections, affecting both pulmonary and wider body systems (Chung and Huh, 2015; Morris *et al.*, 2017) This is particularly a problem for the very old and very young (Christenson *et al.*, 2001; Fortanier *et al.*, 2019).

In the case described above, infection in both grandfather and grandchild could have been prevented with appropriate and timely immunization and prudent social isolation of the infant.

The contribution of nursing leadership in the prevention and management of infection as aligned to the nine leadership dimensions is described below:

Table 8.3. Nursing leadership in the context of vaccination promotion.

Leadership dimension	Application in practice
Inspiring shared purpose Working with others in health and local communities to improve vaccine uptake and preserve health.	Nurses present at a ward round during Max's admission identified immunization in the Emergency Department (ED) as an area for improvement. Working with the local public health nursing team, lead nurse in ED, infectious disease team and the local media developed a plan to communicate the benefits of immunization to the elderly. This involved responding to the public's concerns, and visiting lunch clubs and other social events to establish a visible and approachable relationship with the local community for the benefit of local residents, health services and support agencies.
Evaluating information Use local data on influenza outbreaks to influence preventative actions.	Local epidemiology data on seasonal influenza was used to create short stories in the local media that promoted simple preventative actions. Nurse-led analysis of data and use of appropriate health literacy is used to influence change. Over time, the local population can see cases of influenza dropping.

Continued

Case Study 2: Continued.

Table 8.3. Continued.

Leadership dimension	Application in practice
Connect with services As there is an increase in awareness by the public, opportunities for access to care increase.	Collaboration with the local health protection team identifies requests for vaccinations for other conditions is increasing as the impact of awareness-raising is felt.
Sharing the vision and Influencing for results Creating a common purpose to reduce infection over the winter months produces wide benefits through engagement with the local community.	As a result of Max's family's experience, a desire to share the learning and benefits of vaccination are requested. The local IPC nursing team work with the family to create a personal story which is used to shape local immunization programmes with short-, medium-, and long-term goals. The IPC team work with other local agencies to influence acceptance of this and buy in to maximize impact. 'Ask Your Nurse' becomes a popular local campaign that uses the principles of 'making every contact count' to promote health and reduce infection during the winter months with vaccination.

Conclusion

This chapter has described the current status of nursing leadership associated with AMS. Strategic and clinical leadership are both required to deliver reductions in AMR; however, the journey is in its infancy as evidence and literature builds to support nurses and midwives in these roles. Competency development (Courtenay *et al.*, 2018; Courtenay *et al.*, 2019) will inform future education of the profession and incentivize the creation of further evidence as time evolves. As nurses embrace their position as leaders (NHS Leadership Academy, 2019), nursing leadership will inspire others within the healthcare community to lead with care, purpose and a shared vision to combat AMR. Nurses will bring their assets of knowledge of infection prevention, treatment and follow-up through engagement with other healthcare workers. As recognized, trusted members of the healthcare team, they can bring a voice to the efforts to reduce multidrug resistance through their practice. Nursing leadership by its very nature is holistic (Clarke and Bleich, 2018). The current crisis of increasing multidrug resistance calls for nurses' holistic approach.

Key Points

1. With the emergence of AMS programmes there is an opportunity for clinical nursing leadership.

2. Nurses need to be aware of how they can contribute to AMS programmes and how they can exert leadership to increase public education and awareness of AMR.

3. Core daily nursing functions reflect the principles of stewardship practice.

4. Nurse leaders should proactively engage with other healthcare leaders during policy development processes to identify how and where they should demonstrate their support to stewardship interventions and actions, as well as clarifying the nursing contribution within them.

5. The implementation of stewardship principles and achievement of stewardship goals entails the involvement of nurses in the stewardship process in all stages of patient care.

Further Reading

Department of Health (2014) *Antimicrobial Resistance (AMR) Systems Map*. Available at: https://assets.publishing.service.gov.uk/government/uploads/system/uploads/attachment_data/file/387746/Microbial_Maps.pdf (accessed 10 July 2019).

Lipsitch, M. and Siber, R. (2016) How can vaccines contribute to solving the antimicrobial resistance problem? *mBio* 7(3), e00428–16. Available at: https://mbio.asm.org/content/7/3/e00428-16 (accessed 18 July 2019).

Littman, J. (2015) The ethical significance of antimicrobial resistance. *Public Health Ethics* 8(3), 209–224.

National Institute for Health and Care Excellence (2018) Flu vaccination: increasing uptake. Available at: https://www.nice.org.uk/guidance/ng103 (accessed: 10 July 2019).

O'Neil, J. (2016) Vaccines and alternative approaches: reducing our dependence on antimicrobials. *Review on Antimicrobial Resistance*. Available at: https://amr-review.org/sites/default/files/Vaccines%20and%20alternatives_v4_LR.pdf (accessed 10 July 2019).

RCN (Royal College of Nursing) (2014) *Antimicrobial Resistance – RCN Guidance for the Role and Contribution of Nursing*. RCN, London.

Royal College of Nursing Immunisation (n.d.). Available at: https://www.rcn.org.uk/clinical-topics/public-health/immunisation (accessed 10 July 2019).

The Joint Commission (2017) Antimicrobial stewardship standard (MM.09.01.01). Available at: https://www.jointcommission.org/assets/1/6/New_Antimicrobial_Stewardship_Standard.pdf (accessed 28 September 2019).

References

Allen, D. (2018) The invisible work of nurses. Available at: https://theinvisiblework-ofnurses.co.uk/ (accessed 3 August 2019).

All-Party Parliamentary Group (APPG) on Global Health (2016) Triple impact – how developing nursing will improve health, promote gender equality and support economic growth. Available at https://www.who.int/hrh/com-heeg/digital-APPG_triple-impact.pdf?ua=1 (accessed 3 August 2019).

ANA/CDC (2017) Redefining the antibiotic stewardship team: recommendations from the American Nurses Association/Centers for Disease Control and Prevention Workgroup on the role of registered nurses in hospital antibiotic stewardship practices. *Nursing World*. Available at: https://www.cdc.gov/antibiotic-use/healthcare/pdfs/ANA-CDC-whitepaper.pdf (accessed 28 September 2019).

Bosanquet, J. and Singleton, S. (2014) A burning platform: maximising the nursing contribution to the antimicrobial resistance challenge. Available at: https://publichealthmatters.blog.gov.uk/2014/09/18/a-burning-platform-maximising-the-nursing-contribution-to-the-antimicrobial-resistance-challenge/ (accessed 3 August 2019).

Castro-Sánchez, E. (2018) European Commission Guidelines for the prudent use of antimicrobials in human health: a missed opportunity to embrace nursing participation in stewardship. *Clinical Microbiology and Infection* 24, 914–915.

Castro-Sánchez, E., Gilchrist, M., McEwen, J., Smith, M. and Holmes, A. (2017) Antimicrobial stewardship: widening the collaborative approach. *Journal of Antimicrobial Stewardship* 1(1), 29–37.

Christenson, B., Lundbergh, P., Hedlund, J. and Ortqvist, A. (2001) Effects of a large-scale intervention with influenza and 23-valent pneumococcal vaccines in adults aged 65 years or older: a prospective study. *The Lancet* 357(9261), 1008–1011.

Chung, D. and Huh, K. (2015) Novel pandemic influenza A (H1N1) and community-associated methicillin-resistant Staphylococcus aureus pneumonia. *Expert Review of Anti-infective Therapy* 13(2), 197–207. DOI: 10.1586/14787210.2015.999668

Clarke, P. and Bleich, M. (2018) Holistic leadership – nursing's unique contribution to healthcare. *Nursing Science Quarterly* 31(2), 134–138.

Collins, A. (2008) Preventing health care-associated infections.. In: Hughes, R. (ed.) *Patient Safety and Quality: an Evidence-based Handbook for Nurses*. Agency for Healthcare Research and Quality, Rockville, Maryland. Available at: https://www.ncbi.nlm.nih.gov/books/NBK2683/ (accessed 18 July 2019).

Courtenay, M., Lim, R., Castro-Sánchez, E., Deslands, R., Hodson, K., *et al*. (2018) Development of consensus-based national antimicrobial stewardship competencies for UK undergraduate healthcare professional education. *Journal of Hospital Infection* 100, 245–256.

Courtenay, M., Castro-Sánchez, E., Gallagher, R., McEwen, J., Bulabula, A.N.H., *et al*. (in press) Development of consensus-based international antimicrobial stewardship competencies for undergraduate nurse education. *Journal of Hospital Infection*. Available at: https://www.sciencedirect.com/science/article/pii/S0195670119303135 (accessed 28 September 2019).

Davies, S.C. (2019) Annual Report of the Chief Medical Officer: *Health, Our Global Asset – Partnering for Progress*. Department of Health and Social Care.

Fortanier, A.C., Venekamp, R.P., Boonacker, C.W.B., Hak, E., Schilder, A.G.M., *et al*. (2019) Pneumococcal conjugate vaccines for preventing acute otitis media in children. *Cochrane Database of Systematic Review* 28(5). DOI: 10.1002/14651858.CD001480.pub5

Goff, D.A., Kullar, R., Goldstein, E.J.C., Gilchrist, M., Nathwani, D. (2017) A global call from five countries to collaborate in antibiotic stewardship: united we succeed, divided we might fail. *The Lancet* 17(2), e56–e63.

Greenwood, B. (2014) The contribution of vaccination to global health: past, present and future. Philosophical transactions of the Royal Society of London. Series B, *Biological Sciences*, 369(1645), 20130433. Available at: https://www.ncbi. nlm.nih.gov/pmc/articles/PMC4024226/ (accessed 18 July 2019).

ICN (International Council of Nurses) (2017) Position statement. *Antimicrobial Resistance*. Available at: https://www.icn.ch/sites/default/files/inline-files/ICN_ PS_Antimicrobial_resistance.pdf (accessed 3 August 2019).

ICN (2019) Nursing definitions. Available at: https://www.icn.ch/nursing-policy/ nursing-definitions (accessed 10 July 2019).

Jump, R.L.P., Gaur, S., Katz, M.J., Crnich, C.J., Dumyati, G., *et al*. (2017) Infection Advisory Committee for AMDA – The Society for Post-acute and Long-term Care Medicine. Template for an antibiotic stewardship policy for post-acute and long-care settings. *Journal of the American Medical Directors Association* 18(11), 913–920.

Morris, D., Clearly, D. and Clarke, S. (2017) Secondary bacterial infections associated with influenza pandemics. Available at: https://www.ncbi.nlm.nih.gov/ pmc/articles/PMC5481322/ (accessed 18 July 2019).

Nathwani, D., Sneddon, J., Patton, A. and Malcolm, W. (2012) Antimicrobial stewardship in Scotland: impact of a national programme. *Antimicrobial Resistance and Infection Control* 1:7. doi: 10.1186/2047-2994-1-7.

NHS Leadership Academy (2019) The nine leadership dimensions. Available at: https://www.leadershipacademy.nhs.uk/resources/healthcare-leadership-model/nine-leadership-dimensions/ (accessed 3 August 2019).

Olans, R.N., Olans, R.D. and DeMaria, A., Jr (2016) The critical role of the staff nurse in antimicrobial stewardship: unrecognized, but already there. *Clinical Infectious Diseases* 62, 84–89.

Olans, R.D., Hausman, N.B. and Olans, R.N. (in press) Nursing and antimicrobial stewardship: past, present, and future. *Infectious Disease Clinics*.

Rout, J. and Brysiewicz, P. (2017) Exploring the role of the ICU nurse in the antimicrobial stewardship team at a private hospital in KwaZulu-Natal, South Africa. *Southern African Journal of Critical Care* 33, 46–50.

RCN (Royal College of Nursing) (2016) *Infection Prevention and Control Commissioning Toolkit*. Guidance and information for nursing and commissioning staff in England. RCN, London.

Shlaes, D.M., Gerding, D.N., John, J.F. Jr, Craig, W.A., Bornstein, D.L. *et al*. (1997) Society for healthcare epidemiology of America and infectious diseases society of America Joint Committee on the prevention of antimicrobial resistance: guidelines for the prevention of antimicrobial resistance in hospitals. *Infectious Control and Hospital Epidemiology* 18: 275–291.

Spellberg, B. (2016) New societal approaches to empowering antibiotic stewardship. *Journal of the American Medical Association* 315(12), 1229–1230.

WHO (World Health Organization) (2017) One health. Available at: http://www. euro.who.int/en/health-topics/disease-prevention/antimicrobial-resistance/ about-amr/one-health (accessed 3 August 2019).

WHO (2018) Nursing and midwifery. Available at: https://www.who.int/news-room/ fact-sheets/detail/nursing-and-midwifery (accessed 3 August 2019).

Index

Note: Page numbers in **bold** type refer to **figures**
Page numbers in *italic* type refer to *tables*
Page numbers followed by 'a' refer to appendices
Page numbers followed by 'b' refer to boxes

CABI – who we are and what we do

This book is published by **CABI**, an international not-for-profit organisation that improves people's lives worldwide by providing information and applying scientific expertise to solve problems in agriculture and the environment.

CABI is also a global publisher producing key scientific publications, including world renowned databases, as well as compendia, books, ebooks and full text electronic resources. We publish content in a wide range of subject areas including: agriculture and crop science / animal and veterinary sciences / ecology and conservation / environmental science / horticulture and plant sciences / human health, food science and nutrition / international development / leisure and tourism.

The profits from CABI's publishing activities enable us to work with farming communities around the world, supporting them as they battle with poor soil, invasive species and pests and diseases, to improve their livelihoods and help provide food for an ever growing population.

CABI is an international intergovernmental organisation, and we gratefully acknowledge the core financial support from our member countries (and lead agencies) including:

 Ministry of Agriculture People's Republic of China  Australian Government Australian Centre for International Agricultural Research  Agriculture and Agri-Food Canada  Ministry of Foreign Affairs of the Netherlands  Schweizerische Eidgenossenschaft Confédération suisse Confederazione Svizzera Confederaziun svizra / Swiss Agency for Development and Cooperation SDC

Discover more

To read more about CABI's work, please visit: **www.cabi.org**

Browse our books at: **www.cabi.org/bookshop**, or explore our online products at: **www.cabi.org/publishing-products**

Interested in writing for CABI? Find our author guidelines here: **www.cabi.org/publishing-products/information-for-authors/**